Monitoring for Conservation and Ecology

Edited by
BARRIE GOLDSMITH
Senior Lecturer, Ecology and Conservation Unit,
University College, London

D0321478

CHAPMAN AND HALL

LONDON • NEW YORK • TOKYO • MELBOURNE • MADRAS

UK	Chapman and Hall, 2–6 Boundary Row, London SE1 8HN
USA	Chapman and Hall, 29 West 35th Street, New York NY10001
JAPAN	Chapman and Hall Japan, Thomson Publishing Japan, Hirakawacho Nemoto Building, 7F, 1-7-11 Hirakawa-cho, Chiyoda-ku, Tokyo 102
AUSTRALIA	Chapman and Hall Australia, Thomas Nelson Australia, 102 Dodds Street, South Melbourne, Victoria 3205
INDIA	Chapman and Hall India, R. Seshadri, 32 Second Main Road, CIT East, Madras 600 035

First edition 1991

© 1991 Chapman and Hall

Typeset in 10/12pt Sabon by EJS Chemical Composition, Bath, Avon
Printed in Great Britain by T.J. Press (Padstow) Ltd, Padstow, Cornwall

ISBN 0 412 35590 6 (HB)
0 412 35600 7 (PB)

British Library Cataloguing in Publication Data

Monitoring for conversation and ecology.
1. Ecology and conservation
I. Goldsmith, Barrie
574.50723

ISBN 0–412–35590–6
ISBN 0–412–35600–7 pbk

Library of Congress Cataloguing-in-Publication Data

available

Contents

viii   *Contents*

Preface

Monitoring has become fashionable. Business now talks about monitoring its activities, efficiency, costs and profits. The National Health Service is monitoring general practices and hospitals; it is keen to have more information about efficiency and the duration of stay of patients in different hospitals undergoing different types of treatment. These activities are usually carried out in relation to specific objectives with the aim of making activities more cost effective and competitive. Does the same apply in biology, ecology and nature conservation? Or, are we still enjoying conducting field surveys for the fun of it, at best with rather vague objectives and saying to our colleagues that we do our work because we need to know what is there? This book is an opportunity to consider some of the reasons why monitoring is important, how it differs from survey, how it may be able to answer specific questions and help with site management or problem solving. It will explore some of the taxa that are suitable for recording and how you may actually set about doing it. It is not intended as a catalogue of techniques but we will in each chapter give you sources of material so that with the minimum of effort you will be able to proceed with an efficient, relevant and not too time-consuming monitoring programme. Some of the points that you need to consider before starting are also set down in the synthesis at the end of the book.

Thus, if you already know why you want to monitor and what taxa are ideally suited to your particular circumstances, you may be able to go to the relevant chapters and then refer to the checklist. However, we have endeavoured to compile this book in such a way that all the chapters are readable and support each other so that students and those practising ecology, nature conservation or other land-based sciences will be able to benefit from reading the whole book. This book does not include evaluation which is often closely related to monitoring and has been dealt with elsewhere (Goldsmith 1983) and is itself closely related to the selection of sites for protection which is also discussed elsewhere (Goldsmith 1990).

The idea for the book came from a meeting between staff of the Ecology and Conservation Unit at University College, London and from the Nature Conservancy Council when they sat down to plan a series of one-week modules for both postgraduate students and fee-paying outsiders. Monitoring was suggested as a topic that was academic and mind-stretching, needed by conservationists and could be tailored to about ten half-day slots. A series of ideal speakers was drawn up by the editor and they all graciously agreed to lecture. There was much discussion between them, and as morning speakers mixed with afternoon ones over a pint at lunchtime the idea of this book was conceived. The students were invaluable in making suggestions and criticising the material that was offered. Fortunately most of the original speakers have become authors of chapters and there are one or two additional well-known names that we have subsequently added.

One key distinction that all contributors wanted to make was the difference between survey and monitoring. It was consistently argued that survey was what we have all been doing for years; it is an inventory; it is static in its background philosophy and it is usually done once only. On the other hand monitoring is purpose orientated; it tells us how something(s) is/are changing; it is repeated at regular intervals and it often provides the baseline for recording possible change in the future. Thus it is dynamic in philosophy, purpose orientated and needs to be critically examined prior to implementation. Hopefully after reading this book you will be better equipped to design and conduct a monitoring programme and, in some circumstances, better able to find out whether you are able to achieve your management goal or other objectives.

One issue where there has been considerable recent attention has been atmospheric change, the greenhouse effect and global warming. This is a huge subject where the recording of climatic and atmospheric information is beyond the scope of this book, but more and more ecologists and conservationists are becoming concerned about the effects of variables such as sea-level rise on coastal and wetland ecosystems. We have included how these systems can be monitored by aerial photography, remote sensing and ground truth; how to design a recording system, which taxa to use, and how to decide on priorities and implement a worthwhile project.

I have been involved in two such monitoring projects, one on an internationally important wetland called s'Albufera on Mallorca which involved and was largely funded by Earthwatch. We discovered that the site had a history of drainage and various attempts at cultivation. Thus everything that was seen on the ground, so to speak, was still adjusting or 'relaxing' to past land-use.

This can be seen as an opportunity rather than a problem in that this wetland has a well-documented history, and with certain conditions and combinations of species can be explained in terms of past management. However the problems of using such sites for monitoring long-term and the response to slight changes in parameters such as sea-level are fraught with difficulty. The local staff are also continuously changing current management to make the site better for wildlife which has severe implications for a monitoring exercise!

The other project that I am involved with is being conducted for the World Wide Fund for Nature (WWF) on Prespa National Park in northern Greece. Our group were asked by the Greek authorities to recommend a monitoring programme and we were keen to oblige. It was essential to clarify first the objectives of the monitoring exercise and available resources in terms of skills (for example ecological expertise) and funding. A social problem exists in that the park lies on the Albanian border where it is very difficult to retain specialists due to its isolation. Monitoring overseas for an agency such as WWF was the subject of one of the components of the taught module and this case-study forms Chapter 11 in this book.

Each chapter is designed to stand alone or be read in conjunction with others. Each follows on conceptually from the preceding chapter and in turn leads on to the subsequent one. For some topics it may be appropriate to read a particular combination of chapters and these permutations are indicated below. This book comprises two introductory chapters, the use of aerial photography and remote sensing, monitoring four different groups of organisms, national and county-scale recording schemes, a synthesis involving an application on an RSPB reserve, monitoring overseas, a discussion of the problems involved in the use of index numbers and a North American application of monitoring.

Dr John Hellawell is the monitoring specialist within the Nature Conservancy Council and he provides the introduction (Chapter 1) which includes definitions and a discussion of the importance of defining one's objectives at the outset of any project. Dr Michael Usher in Chapter 2 focuses on the purpose of the monitoring, choice of method, methods of analysis, interpretation of data and results and decisions about when to terminate a monitoring programme.

Dr Jonathan Budd in Chapter 3 discusses both aerial photography and the use of satellite imagery and makes a useful comparison between them. These techniques are particularly useful for recording habitat and landscape change. This chapter might be read in conjunction with Chapter 5.

The chapters dealing with selected groups of organisms commence with Dr Mike Hutchings dealing with plant populations. Usually rarities are given high priority and his examples include an account of his important work on orchid populations, especially the early spider orchid. Chapter 5 deals with vegetation at the community level although we are reminded that monitoring vegetation types can be rather insensitive and therefore not necessarily recommended. This chapter links with Chapters 2, 3, and 11.

A chapter dealing with monitoring butterflies follows. Dr Ernie Pollard describes the technique that he first used in 1973 and which is now being applied to 58 species on about 80 sites. The information gained can be used as a surrogate for the conservation value or 'health' of agricultural land and is referred to again in Chapter 11. Dr Stephen Baillie follows with Chapter 7 which describes the scheme used by the British Trust for Ornithology for censusing birds. Their schemes are also widely used and much respected overseas. He also describes a scheme used in the United States. Further comments on the use of index values, as used in Chapters 6 and 7, appear in Chapter 12.

Dr Paul Harding in Chapter 8 discusses mapping schemes which operate for 6000 different taxa mostly at national level. Whilst not strictly monitoring, their results are extremely useful to ecologists and nature conservationists. Some mapping schemes are also discussed in Chapter 5.

Claire Appleby in Chapter 9 considers techniques of data handling at the county level. Her experience comes mainly from Wiltshire but her recommendations and advice are appropriate to recording schemes for any area of county size (50 × 50 km).

Kevin Roberts is a practical reserve manager and only monitors when he considers that there is a good reason. Chapter 10 candidly explains his philosophy and approach and conservationists in his position may wish to start reading here and then follow with other chapters, especially 2, 4, 5, 7, 11 and 12.

My second Chapter (11) deals with the selection of priorities on relatively unfamiliar sites. I have chosen to describe a study carried out for the WWF at Prespa National Park in northern Greece. It took some time to decide on the criteria to select the key features and key taxa and I hope that this account will save the time of others elsewhere. Chapters, 1, 2, 3, 5, 6, and 9 are particularly relevant as supporting chapters.

Dr Terry Crawford's Chapter 12 deals with the problems of using index numbers for any taxonomic group. This is particularly relevant to Chapters 6, 7 and 13. The final chapter is from North America and brings together some topics referred to earlier. Dr Paul Keddy is a wetland specialist and he discusses the selection of state variables, the relationship between prediction and monitoring and concludes with the possibility of formulating a national state of the environment index for Canada (see Chapter 13).

By definition a monitoring programme must have clear *objectives*. Most that we have encountered tend to fall into one or other of the following groups which are not in any particular order and not necessarily exclusive:

1. to record long-term environmental change and its ecological effects;
2. to record response to a changing management factor such as drainage, grazing, use of herbicides, or agricultural intensification;
3. to record rate of change such as decline of a rare species or habitat, e.g. rate of loss of chalk grassland or monitoring of orchid populations;
4. to determine the effectiveness of a form of management (e.g. grazing versus cutting), or to determine and reduce costs.

When one selects a monitoring programme there are a variety of problems or *constraints* that have to be taken into consideration, such as:

1. long-term directional changes being complicated by 'normal' seasonal and annual cycles;
2. irregular and natural fluctuations which may mask the features of interest (see Chapter 2);
3. ecosystems and species adjusting to current and recent management especially when discontinued or changed, e.g. changing grazing pressure (these are sometimes brought about by social changes);
4. adjustment or 'relaxation' to historical changes, such as drainage, fire, clearcutting of forests;
5. successional change, which can be either directional or cyclical.

I hope this is not discouraging but it is better to consider these points at an early stage rather than find that the reader has started his monitoring without considering them. Very early on he also needs to consider his *priorities* because money, time and expertise are never limitless! In our Prespa project I decided that features of international importance were more important than those of national importance and that those of national importance were more important than those of regional importance. I then gave priority to monitoring habitats and species that were important on the widest scale (see Chapter 11).

Constraints are often readily apparent whereas *opportunities* are more difficult to identify! One opportunity occurs where one or more species are particularly sensitive to a physical or chemical factor of particular importance. It is then possible to monitor the abundance or performance of that species in order to record changes in the key factor. For example, *Cladium mariscus* in wetlands could be used to monitor salinity as the species disappears when salinity exceeds a fairly low level. The problem of this approach, a kind of bioassay, is that the performance of the species varies according to the season (time of year), stage of growth, level of grazing pressure, intensity of competition from other species, stage of succession, etc. (see Chapter 2).

Some 'rules' for the *selection of sites* also have to be considered.

1. They must have security. Often nature reserves are suitable because they are wardened and management is less likely to change without notice. In many circumstances however they cannot be used, for example, to study the effects of agrochemicals or current forestry practices.
2. They, or their habitats or species, must be sensitive to the feature of concern or interest.
3. The local people must support the project and be able to provide help with it. It is also extremely useful to have local knowledge (or at least experts

from a local university) with expertise about species identification, migration, habitat management and ecological processes.
4. It is desirable to have information about past management and it is an advantage if there is a good archive locally.
5. The site should be sufficiently resilient to survive the expected duration of the project and the monitoring which will be conducted. Some of our monitoring sites on the wetland of s'Albufera, Majorca, disappeared within 12 months.
6. The site should also be sufficiently resilient to accommodate the sampling activity which will be necessary. Some alpine, coastal dune and wetland areas are not very resilient and plans then have to be made accordingly.

If you have thought about all these topics you are now ready to go to the chapter of your choice but we would prefer you to read them all in order to be able to put together all the pieces of the jigsaw.

REFERENCES

Goldsmith, F.B. (1983) Evaluating Nature, pp. 233–46 in *Conservation in Perspective*, eds. A. Warren and F.B. Goldsmith. Wiley, Chichester.
Goldsmith, F.B. (1990) 'The Selection of Protected Areas' in *Scientific Management of Temperate Communities for Conservation*, eds. I. Spellerberg, F.B. Goldsmith and M.G. Morris. British Ecological Society Symposium no. 31. Blackwell, Oxford.

Contributors

CLAIRE E. APPLEBY The World Conservation Monitoring Centre, Cambridge, UK, formerly at Wiltshire Biological Records Centre, Devizes, Wiltshire, and the Institute for Terrestrial Ecology, Monks Wood Experimental Station, Huntingdon, Cambridgeshire, UK

STEPHEN R. BAILLIE British Trust for Ornithology, Tring, Hertfordshire, UK

JONATHAN T.C. BUDD Nature Conservancy Council, Peterborough, Cambridgeshire, UK

TERENCE J. CRAWFORD Department of Biology, University of York, York, UK

BARRIE GOLDSMITH Ecology and Conservation Unit, University College London, London, UK

PAUL T. HARDING Environmental Information Centre, Institute for Terrestrial Ecology, Monks Wood Experimental Station, Huntingdon, Cambridgeshire, UK

JOHN M. HELLAWELL Nature Conservancy Council, Peterborough, Cambridgeshire, UK

MICHAEL J. HUTCHINGS School of Biological Sciences, University of Sussex, Brighton, Sussex, UK

PAUL A. KEDDY Department of Biology and Institute for Research on Environment and Economy, University of Ottawa, Ottawa, Ontario, Canada

ERNEST POLLARD formerly at the Environmental Information Centre, Institute for Terrestrial Ecology, Monks Wood Experimental Station, Huntingdon, Cambridgeshire, UK

KEVIN A. ROBERTS Royal Society for the Protection of Birds, Hoddesdon, Hertfordshire, UK

MICHAEL B. USHER Department of Biology, University of York, York, UK

– 1

Development of a rationale for monitoring

JOHN M. HELLAWELL

1.1 INTRODUCTION

'Monitoring' has become an omnibus term and is sometimes applied, almost indiscriminately, to a range of disparate activities. Amongst these one may include, for example, attempts at describing prevailing environmental conditions; the occurrence, distribution and intensity of pollution; the status of ecological communities or populations of species; or simply providing a watching brief on the countryside at large.

Since 'monitoring' is a process, not a result, a means to an end rather than an end in itself, it should not be surprising to find that so many kinds are undertaken.

Implicit in the rationale for most monitoring activities is a recognition of the potential for change. One is concerned, therefore, to secure a means of detecting that a change has occurred, of establishing its direction and of measuring its extent or intensity. This stage may prove to be the simpler part of the monitoring process: often it is more difficult to assess the significance of the change which has been encountered. Monitoring schemes, especially those concerned with ecological change, may founder through lack of adequate criteria for significance and even well-established procedures for pollution monitoring may, in reality, be based on largely arbitrary limits of acceptability for given pollutant concentrations.

This chapter will, above all, stress the importance of establishing clearly-defined objectives in order to ensure the development of a successful monitoring strategy.

1.2 DEFINITIONS

Although 'monitoring' is still often used in a very broad sense, more recently it has acquired a stricter definition (Hellawell 1978) and there is evidence that this is becoming widely accepted. It will be seen from subsequent sections that insistence on greater precision is not merely a question of semantics: the design of a monitoring strategy is greatly assisted by the adoption of clearer definitions. These are as follows.

2 Development of a rationale for monitoring

(a) Survey
An exercise in which a set of qualitative or quantitative observations are made, usually by means of a standardised procedure and within a restricted period of time, but without any preconception of what the findings ought to be.

(b) Surveillance
An extended programme of surveys, undertaken in order to provide a time series, to ascertain the variability and/or range of states or values which might be encountered over time (but again without preconceptions of what these might be).

(c) Monitoring
Intermittent (regular or irregular) surveillance carried out in order to ascertain the extent of compliance with a predetermined standard or the degree of deviation from an expected norm.

It will be seen that while surveys and surveillance are, to a large extent, open-ended, the institution of a monitoring programme imposes a considerable degree of discipline since the standard or norm has to be defined or formulated, however vaguely, before the programme can be implemented.

Examples of the sorts of standards which might be used in ecological monitoring include the size of an animal population; the biomass of vegetation; growth, production or recruitment rates; checklists of species or species richness; community diversity indices; the extent or structure (mosaic or diversity) of habitats; vegetation (phytosociological) classifications, and the presence or absence of indicator species. In fact, almost any appropriate measure could be employed as the yardstick for monitoring. As will be noted, the list includes dynamic processes (e.g. production) as well as static measures (species checklists). Monitoring is undertaken to ascertain whether the prevailing conditions (physiological, behavioural, ecological or environmental) match the previously defined standards or norms, expressed perhaps as acceptable minima or maxima, or that they lie within certain defined limits.

It is evident that survey and surveillance differ from true monitoring, as defined above and as employed in this chapter, in both the intention for instituting them and in the likely consequences of undertaking them. The former are carried out without preconception, at least without explicit preconceptions, in an attempt to acquire information as objectively as possible. If one already knew the results, it would be difficult to justify the expenditure of resources merely to confirm them! Monitoring, on the other hand, presupposes that one already has an idea, however vague, of the results which one expects to obtain. Even vague concepts, for example whether a site matches up to an idealised or typical description, can be pressed into service for monitoring. In such cases it may not be possible to say much more than the prevailing conditions are atypical: more intensive investigations may then be required to justify and qualify the conclusion.

This aspect leads to a consideration of the second major difference between monitoring and survey or surveillance which is that monitoring is intrinsically purposeful. Of course, one conducts surveys with a purpose, that is, to enhance knowledge or, at least, satisfy curiosity. But in monitoring one is concerned with setting limits, however arbitrary, and in deciding what action may be necessary when monitoring reveals that the current situation is wide of the target. Sometimes this aspect is not considered early enough in the development of a strategy. It is arguable that it is pointless to deploy effort in monitoring a situation over which one has no effective control or for which no response would be required.

1.3 WHY MONITOR?

The reasons for instituting a monitoring programme are legion but may be classified into three general categories. These are:

1. assessing the effectiveness of policy or legislation;
2. regulatory (performance or audit function);
3. detecting incipient change ('early warning').

It should be emphasised that these categories are not mutually exclusive since, for example, legislation may place statutory duties on a regulatory authority to anticipate change as in the case of the Nature Conservancy Council Act (1973) which, in Section 1(5), requires the Council 'to take account, as appropriate, of actual or possible ecological changes'.

One activity for which monitoring is not relevant is research, although one may encounter examples where this is claimed. For example, it might be explained that a management practice at a site, such as grazing, has been radically changed and the consequences are being 'monitored'. The definition employed in this chapter does not admit of such a usage since research is undertaken in order to ascertain the unknown or uncertain outcome. If the conclusion is already known, the research is unnecessary! In research the provision of adequate controls is also important; in the example cited above it would be necessary to continue to manage part of the site as previously. Otherwise one could not acquire evidence that the outcome was strictly attributable to the change in management. However, if one had changed the management with the intention of securing a particular outcome (that is a standard or norm in the true monitoring sense) it would be correct to monitor progress towards that outcome and when prevailing conditions matched the standard specified, then one could conclude that the purpose had been attained. From then on monitoring could provide assurance that this standard was being maintained.

1.3.1 Monitoring effectiveness of policy or legislation

Legislation is commonly intended to secure the maintenance of a desirable condition or facilitate progress towards such a condition. For example, the preamble to the Rivers (Prevention of Pollution) Act (1951) states that the purpose of the Act is 'to make new provision for maintaining or restoring the wholesomeness of the rivers and other inland or coastal waters of England and Wales ...'.

Provided one has an adequate definition of 'wholesomeness', it is possible to monitor its maintenance or progress towards that state in waters which do not presently comply with the necessary standard. Similarly, one could monitor the numbers of Sites of Special Scientific Interest which survive intact, or which continue to retain their conservation interest, as an indicator of the efficiency of the relevant legislation, namely the Wildlife and Countryside Act (1981). Sometimes, monitoring of legislation might be more indirect. Under the provisions of Section 24(1) of the Wildlife and Countryside Act (1981) the Nature Conservancy Council is required to review, quinquennially, the species to be protected and listed in Schedules 5 (animals) and 8 (plants) although it may do this, as necessary, at any time. In order to ensure that rare and/or endangered species appear in the schedules it will be necessary to monitor the status of all populations, although the method might be two-tier, giving minimal attention to those species which are common or ubiquitous at the present time.

1.3.2 Regulatory monitoring (performance or audit)

Much of the day-to-day monitoring for management purposes is to be included under this heading. Pollution-regulating authorities are routinely involved in monitoring compliance with standards of water effluent quality, air-pollutant emissions, noise and radioactivity. Similar activities are necessary in areas of health and safety, food hygiene, medical care and just about every other area of human activity.

The ecological audit function included under this general heading is usually concerned with such activities as checking that management agreements for maintaining the quality of sites are effective. In the present context, the emphasis would be on the site, not simply on policing compliance with the terms of an agreement, although such checks are clearly advisable. But for purposes of monitoring, the essential question has to be that of the maintenance (or attainment) of the desired condition: it is quite possible that adhering strictly to the terms of an agreement might not be sufficient if the proposed management regime were to be flawed or if other factors were at work.

1.3.3 Detection of incipient change

The last point makes a convenient link with the third category, the use of monitoring to detect incipient (and, usually, undesirable) change. This is often the area of greatest interest to ecologists and conservationists who are invariably concerned about the external pressures on wildlife and threats to its continued survival.

In ecosystems there are broadly three kinds of intrinsic changes: stochastic, successional and cyclical. Stochastic changes are, by definition, unpredictable and therefore cannot be anticipated by monitoring schemes. Stochastic changes may be associated with severe climatic events such as floods, droughts, fires and epidemics of diseases or parasites. The adverse effects of severe winters on the survival of bird populations is well documented (Lack 1966), as is the beneficial effect of warm dry summers on the recruitment of fish stocks (Craig 1980). Severe events may, ultimately, have beneficial effects, for example in breaking seed dormancy, but more often the immediate consequences are detrimental. However, the innate resilience of many biological systems facilitates a partial or even complete recovery. Here, monitoring can play a part in following progress towards recovery and in confirming that recovery has occurred. Successional changes may be extremely slow, giving an impression of stability but ultimately resulting in significant change. The characteristics of ecological succession have been reviewed by Odum (1969). Some successional changes may be reversed or halted by appropriate management practices, but succession is a normal ecological process which results in a gradual change of communities and the disappearance of species. Cyclical changes may, in the short term, be quite dramatic in their effects, but usually contribute to the indefinite persistence of populations. Examples of cyclical changes include predator–prey and density-dependent interactions (MacLulich 1937; Elton 1942; Wynne-Edwards 1965).

Although one may classify changes as cyclical, successional and stochastic, in reality all three may occur simultaneously or be superimposed. Cyclical changes may take place within a general successional trend, punctuated from time to time by stochastic events which temporarily obliterate the pattern.

One of the major influences on ecosystems is human activity and the purpose of much monitoring is to assess impacts and recovery following amelioration. The principal culprits are industrial and urban development, with their attendant problems of pollution and disturbance, and also trends in agricultural and forestry practice including enhanced use of fertilisers and pesticides.

Human activity has also created habitats of value: it is often these artificially generated and managed systems, such as coppiced woodlands and chalk grassland, with their sub-climax communities, which support the richest floras and faunas. The contribution of monitoring programmes is to provide confirmation of their continued value and early warning of the effects of excessive human or management pressure.

1.4 DESIGN OF MONITORING STRATEGIES

Anyone about to embark on a monitoring programme is faced with a number of questions of which the most immediate are 'what are my objectives?' and hence 'what is to be monitored?' As explained above, this involves the selection of a standard or norm against which changes can be assessed. Once this has been identified, the subsequent questions, 'how?', 'when?', 'how often?' are more readily tackled.

The importance of identifying precise objectives cannot be over-emphasised. All too often programmes have been instituted without, apparently, more than a vague idea of what is the purpose and, one can only presume, in the hope that after the data have been subjected to elaborate statistical analyses, something will, in true Micawber fashion, turn up! Opponents of the view advanced in this chapter occasionally draw attention to the serendipity evident in earlier monitoring programmes where, at their inauguration, their subsequent value in providing evidence of significant changes, could not have been anticipated. It is then argued that this justifies, retrospectively, the collection of all and any data which might be potentially useful in the future. Yet hindsight is no substitute for foresight: every programme which ultimately yielded worthwhile results by data-kleptomania would be greatly outnumbered by many which proved sterile. In most instances, the cost of collecting all possibly relevant data now, in case it proved useful in the future, would be prohibitively high.

In discussing the inherent problems of monitoring strategy design, reference is frequently made to the radio reception analogy of 'signals' and 'noise'. The whole spectrum of environmental or ecological parameters contains enormous amounts of information but most is unintelligible, that is, we can obtain the data but we may not understand their significance. What is required, in terms of the analogy, is the development of a 'tuner' which filters out the 'noise' leaving a clear 'signal'. This signal becomes the indicator of significant change. The quest for reliable 'tuners' and clear 'signals' has exercised the minds of those concerned with monitoring for a considerable time. Several problems are often encountered. Continuing the analogy, we may have a clear, finely-tuned signal on one channel, yet miss vital 'news' on another. Early warning of significant change is clearly highly desirable but early warnings begin as faint signals amidst considerable noise. By the time the signal is loud and clear, the message may be too late!

1.4.1 Selection of key indicators

The selection of the key indicators from which a 'signal' is to be generated depends on the objectives of the monitoring exercise. Where a specific issue is involved, for example the indefinite maintenance of the last surviving stand of a flowering plant, it should prove possible to devise a straightforward measure such as, in this example, the total number of plants, or their reproductive capacity and seedling survival, as a standard. However, in complex habitats or extensive sites, one may need to try to find a single component, the behaviour and responses of which are indicative of the community as a whole. This is a tall order: few species are understood well enough to permit their use as ecological 'litmus-paper'. The complexity of the problem can be seen by reference to Figure 1.1 which provides a conceptual framework of community organisation and indicates the structural or functional measures which might be employed in monitoring at each level. The complexity of ecosystems extends beyond that illustrated in the figure: groups of communities contribute to habitat and site characteristics and these combine to form parts of ecosystems.

The individual organism, ignoring for the moment its taxonomic identity, appears to be the least useful component although its physiological attributes might prove relevant. The latter, together with tissue analyses, are often utilised in pollution monitoring studies. The presence of individual species or, more usually, populations of species is often of great value in assessing habitat or environmental quality when their 'indicator' value is known. Biocoenoses, or groups of species commonly found together, are qualitative attributes of ecosystems which have considerable value: plant species groups have been studied exhaustively by phytosociologists and they form the basis of several habitat classification schemes, including, for example, the National Vegetation Classification (NVC) scheme funded by the Nature Conservancy Council (NCC 1989).

Communities are composed of assemblages of populations and biocoenoses which may be used to derive quantitative measures such as diversity (the distribution of numbers, or biomass, between the various taxa) or production (a dynamic measure, defined as the total biomass elaborated in unit time). Parameters such as these probably qualify as the nearest approach to the ideal single key measure but their measurement is extremely demanding in resource terms.

1.4.2 Selection of an effective approach

There are two basic approaches to monitoring strategies, one of which (the 'serendipity' or 'Micawber' method) has already been alluded to above. It is

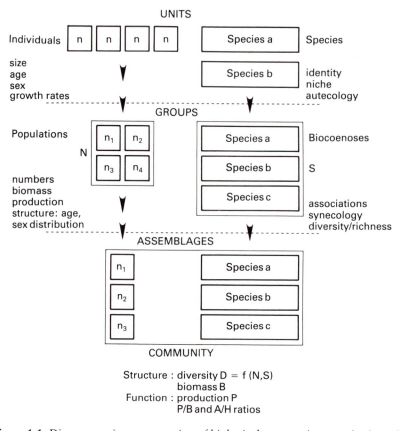

UNITS

| Individuals | n | n | n | n | | Species a | Species |

size
age
sex
growth rates

GROUPS

identity
niche
autecology

Populations			Species a	Biocoenoses
N	n_1	n_2	Species b	S
	n_3	n_4	Species c	

numbers
biomass
production
structure: age,
sex distribution

associations
synecology
diversity/richness

ASSEMBLAGES

n_1	Species a
n_2	Species b
n_3	Species c

COMMUNITY

Structure : diversity $D = f(N,S)$
biomass B
Function : production P
P/B and A/H ratios

Figure 1.1 Diagrammatic representation of biological community organisation with the structural or functional aspects which may be utilised in monitoring change. P = Production; B = Biomass; A = Autotrophic; H = Heterotrophic. (After Hellawell 1977.)

often encountered and the process is illustrated in Figure 1.2. The conduct of field surveys over a period of time generates a considerable quantity of data which, in turn, are subjected to statistical analyses. Sometimes the data are subjected to the whole of the contents of a standard computer statistical package! Not infrequently, the results are of two kinds: either the data are inadequate to meet the rigours of the statistics or the answer is inconclusive. More field surveys are then required in an attempt to remedy the inadequacies. This open-ended monitoring strategy is avoidable, provided that clear objectives are set and a true monitoring yardstick is defined at the outset.

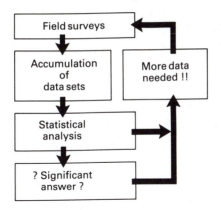

Figure 1.2 Schematic representation of stages in many conventional monitoring programmes, leading to equivocal results and repeated cycles.

The preferred approach is illustrated in Figure 1.3, with particular reference to monitoring of sites. The overall objective is to secure the indefinite maintenance of the conservation interest of the site. A management plan would be drawn up independently: the scientific monitoring is intended to provide confirmation of the site's continued value for conservation. The

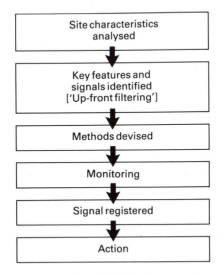

Figure 1.3 Schematic representation of the ideal monitoring strategy, illustrated by reference to a site, in which the signal is determined at the outset and only the monitoring process is repeated, as required.

first stage is to identify key components which are intrinsic to the status of the site. These may be the presence of rare, notable or endangered species, characteristic communities or, more rarely, an important physical measurement such as the level of the water table. This last example is an attractive proposition in many situations since physical measurements are often more readily undertaken than ecological ones. However, such determinants suffer from the primary disadvantage that they are usually only an indirect means of assessing the ecological condition of the site. It is quite possible, for example, for the water table to remain at a constant level and for species to be affected by other factors.

Once key features have been identified, it is necessary to select a 'signal' which will form the basis of the monitoring programme. If the key feature was species richness, a figure would be chosen, below which one would decide that action would have to be taken. Suppose the site normally supported about 150 plant species and surveillance had established that, over a number of years, there had always been at least 135 recorded. One might decide that the action limit would be set at 135 or even 130 species. If the site would still retain conservation interest with fewer species, or with the continued presence of an even smaller set of essential species, then one could adopt this as the yardstick. This part of the process is likely to cause considerable heart-searching and debate. It is essential that the decision is made, even if somewhat arbitrarily. Otherwise, there is considerable danger of getting stuck on the merry-go-round of the system depicted in Figure 1.2. Although a decision is essential it need not be irrevocable: should experience reveal that the original result of the prerequisite filtering was unsound, it can be changed, as necessary. The important point is that some signal, the trigger for action, is determined before the monitoring regime is instituted.

Having determined the signal, the selection of a method for its detection becomes fairly straightforward. At this stage a monitoring 'prescription' is drawn up, specifying the method and frequency of sampling. Experience or resource constraints might suggest an appropriate return-frequency, depending on the likely rate of change. When the predetermined degree of change is detected, the appropriate action is elicited. The subsequent restoration of the required status is also established by application of the same monitoring strategy.

It must be expected that the sensitivity and precision of this approach will vary according to the system under scrutiny, the objectives of the monitoring and the adjustments which can be made as a result of accumulated experience. The use of biological indicators as a technique for monitoring freshwater environments is now well established (Hellawell 1986) but it has taken almost a century of development and refinement. Yet, even here, further advances are likely to occur.

1.4.3 Common pitfalls in project design

The importance of clearly defined objectives may be illustrated by reference to some common pitfalls in the design of monitoring projects. The difficulties often experienced in the use of random sampling for determining parametric statistics in ecology, often a consequence of the highly variable nature of living systems, has engendered a proclivity for the adoption of fixed quadrats in monitoring. This enables one to sample repeatedly the same quadrat over time, noting changes, on the assumption that the microcosm within the quadrat will behave in exactly the same way as the whole habitat. This last element is secured, as far as possible, by selecting an appropriately representative area in which to site the quadrat. The approach is quite adequate for many purposes, provided that any change observed within the quadrat is typical of the site or habitat as a whole. Where the site is managed in a fairly uniform way and the quadrat is subjected to exactly the same treatment as the rest of the habitat, one could assume, quite reasonably, that the fixed quadrat provides a good indication of changes.

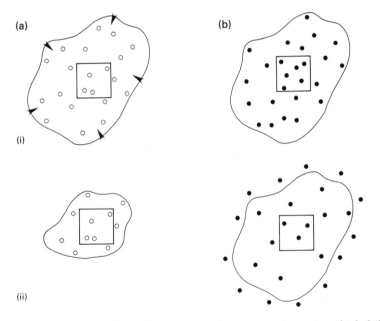

Figure 1.4 (a) Use of fixed quadrat at a central representative point which fails to detect change at periphery of site.
(b) Use of fixed quadrat may suggest decline in population, as indicated by lower density, when total remains constant.
(i) Before; (ii) after.

Reference to Figure 1.4a shows, however, that a quadrat placed in the centre of a shrinking population, stand or habitat may not reveal any change until the situation has become quite critical. The opposite condition is illustrated in Figure 1.4b. Here the selected signal is the population size. A fixed quadrat placed in the centre may reveal a decline in the density of the stand, but the outward spread of the constant population is not detected. One might conclude, incorrectly, that the population was declining. Quadrats, sited in representative areas of a mosaic of communities may be inappropriate if one is intent on detecting changes in their extent. For example, in Figure 1.5a the boundary between two communities shifts in a north-westerly direction. Until there is a pronounced change in the left-hand quadrat, this

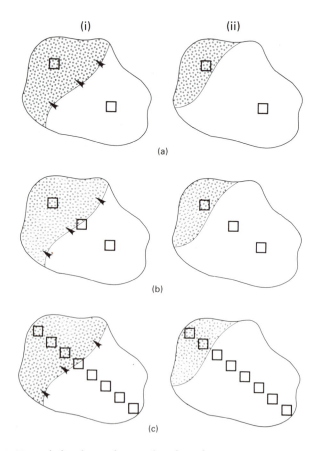

Figure 1.5 Use of fixed quadrats, placed within representative areas of two communities, fails to detect boundary change (a). Extra quadrat (b) on boundary detects change but use of transact (c) provides more information.
(i) Before; (ii) after.

could go undetected. In Figure 1.5b an additional quadrat, placed on the original boundary, would provide an indication of this change. Far better, however, would be the setting up of a transact (or a line of smaller quadrats) across the interface between the two communities (Figure 1.5c). In this way, not only the change but also the rate of change could be measured.

It should be emphasised that the very simple examples described above are intended to illustrate the kinds of problems which might arise from the indiscriminate use of fixed quadrats.

A common concern in monitoring programmes is the establishment of 'baselines', a term which has acquired a variety of meanings, including the condition which prevails when monitoring begins, and the basis from which all future change is assessed. This would be quite acceptable, were it not for the difficulty of knowing how representative was the state of affairs when the programme began. As indicated above, many systems exhibit cyclical or stochastic changes. A period of surveillance is, therefore, desirable to ensure that, when a baseline is established, it does represent a base: for example, the lowest population normally encountered. In Figure 1.6, which shows the erratic fluctuations of a population over time, a programme instituted during the middle period (b) would suggest baseline 1 would be appropriate. If the surveillance had begun earlier, when the population recovered well from a low value (a), baseline 2 would seem to be equally acceptable. This receives confirmation at time c when once more, the population increased again. At time d one would regard the continued downward trend as a cause for concern and action might be required. Only time would reveal whether another baseline was being established, but this hypothetical example illustrates the principle that fixing baselines without a period of surveillance is unwise, and baselines may need to be revised in the light of experience.

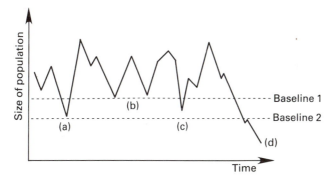

Figure 1.6 Diagrammatic representation of variation in population number with time. Surveillance during period (b) would indicate baseline 1 is appropriate while earlier and later periods, (a), (c) suggest baseline 2 is acceptable. At time (d), action is required.

1.5 CONCLUSION

In this review of the basic principles of monitoring, the need for clear objectives and the design of programmes which meet these, has been emphasised. It may seem like a statement of the obvious but, in the past, many projects have foundered on a failure to do this, and monitoring has acquired a poor reputation. Schemes have incurred high costs and given few tangible benefits.

Scientific research and development, like many spheres of human activity, has its cycles of fashion. The author can remember a period of intense interest in monitoring about two decades ago followed by a progressive decline when disillusionment set in. Many projects were too ambitious and the results were often equivocal. It was also obvious that such work does not lend itself to acquiring an extensive list of publications nor academic, or other, honours.

One now senses a revival of interest in monitoring, promoted, no doubt, by greater public awareness of environmental issues. The need for information on the possible effects of anticipated global warming, the loss of the ozone layer and the other consequences of human development and activity is creating a climate in which monitoring issues are active once more. It is to be hoped that a more effective approach to the development of monitoring programmes will better meet the challenges which face us.

REFERENCES

Craig, J.F. (1980) Growth and production of the 1955–1972 cohorts of perch, *Perca fluviatilis* L. in Windermere. *Journal of Animal Ecology* **49**, 291–315.

Elton, C. (1942) *Voles, mice and lemmings: problems in population dynamics.* Oxford University Press, London.

Hellawell, J.M. (1977) Change in natural and managed ecosystems: detection, measurement and assessment. *Proceedings of the Royal Society. London B*: **197**, 31–57.

Hellawell, J.M. (1978) *Biological surveillance of rivers.* Medmenham and Stevenage: Water Research Centre.

Hellawell, J.M. (1986) *Biological indicators of freshwater pollution and environmental management.* Elsevier Applied Science Publisher, London.

Lack, D.L. (1966) *Population studies of birds.* Clarendon Press, Oxford.

MacLulich, D.A. (1937) *Fluctuations in the numbers of the varying hare* (Lepus americanus). University of Toronto Studies, Biological Series No. 43.

NCC (1989) *Guidelines for selection of biological SSSIs.* Nature Conservancy Council, Peterborough.

Odum, E.P. (1969) The strategy of ecosystem development. *Science*, **164**, 262–70.

Wynne-Edwards, V.C. (1965) Self-regulating systems in populations of animals. *Science*, **147**, 1543–8.

2

Scientific requirements of a monitoring programme

MICHAEL B. USHER

2.1 INTRODUCTION

If a monitoring programme is being planned, there are five basic questions that need to be asked and answered. Each question is important and should be answered before any monitoring begins; they essentially form a sequential set because a satisfactory answer to any individual question cannot be given until all questions higher on the list have been answered. The questions are:

1. **Purpose:** what is the aim of monitoring?
2. **Method:** how can this aim be achieved?
3. **Analysis:** how are the data, which will be collected periodically, to be handled?
4. **Interpretation:** what might the data mean?
5. **Fulfilment:** when will the aim have been achieved?

The aim of this chapter is to follow through these questions, discussing aspects of each that relate to the data being collected and the interpretations likely to be placed on these data. Many of the subsequent chapters address specific issues raised by these five broader questions.

2.2 PURPOSE

To some extent, the purpose will have been defined before consideration of the monitoring begins. The aim of the monitoring might be to estimate the annual size of a population: for example, how many bee orchids, *Ophrys apifera*, are flowering in the population in a nature reserve? (For many plants, however, the number of flowering spikes is a poor indicator of the population size.) It might equally be to estimate part of the population, such as the number of territories of the great tit, *Parus major*, in a farm woodland. It might be that the monitoring aims to establish a concentration by, for example, sampling air for its SO_2 concentration or a river for its NO_3 concentration or BOD (biological oxygen demand). Alternatively, monitoring may aim to determine whether or not some event takes place, for example, much monitoring of the biota in river systems is designed to detect, a posteriori, if a pollution incident has occurred.

Two aspects of the monitoring should also be linked to its purpose. First, there is the *intensity* of monitoring; second, there is the *frequency* of monitoring.

2.2.1 Intensity of monitoring

Intensity is important if an estimate of the reliability of monitoring is required. It may be possible to count every bee orchid flowering in the nature reserve if the population is small both in number and area, but for many populations an estimate of the mean number of orchids per unit of area, together with a standard error of that mean, is required. Similarly for concentration, a mean and a measure of its reliability is required. Using the example of a concentration, the true mean μ is estimated by a sample mean m, and the true standard deviation σ by the sample standard deviation s. The standard error of the mean is given by

$$s/\sqrt{n}$$

where n is the number of observations in the sample (such simple statistical calculations are described in all introductory texts to statistics, e.g. Bailey (1981), Campbell (1974) and Parker (1979)). Confidence limits for μ are $m \pm ts/\sqrt{n}$, where t is Student's t with some defined probability level (2-tailed test); the confidence limits imply an interval around m within which the true, but unknown, value of μ may be regarded as lying with a given degree of certainty. The true mean, μ, is the feature of the monitoring programme that is of particular interest.

Taking arbitrary standard deviations of one, two and four ($s = 1, 2$ and 4), Figure 2.1 shows how the 95% confidence limits become closer to the mean as the size of the sample increases; very small sample sizes have very wide confidence intervals whereas larger sample sizes have narrower confidence intervals. The data in Table 2.1 show how the confidence limits are reduced as the sample size is successively doubled. Although greater precision can be obtained by taking very large samples, in reality there has to be a trade-off between the accuracy of the estimate of μ and the cost of taking and analysing the samples.

There are two problems with the practical application of the approach outlined in Figure 2.1 and Table 2.1. First, it was assumed that the samples were normally distributed, or at least approximately so. This is probably a reasonable assumption for estimates of a concentration but it is certainly not reasonable for counts of a species in quadrats, etc. These are at best likely to be Poisson (randomly) distributed, but more likely to show some kind of

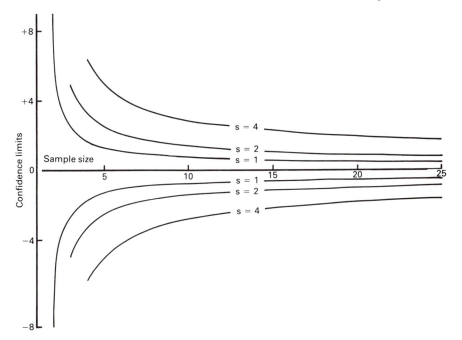

Figure 2.1 The decrease in the 95% confidence limits with increasing sample size. The lines are drawn for different values of the standard deviation (*s*); Student's *t* has been used in calculating the confidence limits. Note that as the sample size approaches infinity, the confidence limits all approach zero.

Table 2.1 95% confidence limits for samples of varying size assuming a standard deviation of one (*s* = 1) and a Student's *t* distribution

Sample size	95% confidence limits	Percentage decrease
2	± 8.984	
		82
4	± 1.591	
		47
8	± 0.836	
		36
16	± 0.533	
		32
32	± 0.361	
		31
64	± 0.250	
		30
128	± 0.175	

aggregated distribution. The importance of using logarithms of counts is outlined later (see Chapter 12). Second, the size of the standard deviation is unknown before samples are taken. It may, therefore, be wise to take a preliminary sample so as to gauge the magnitude of the standard deviation, and on the basis of that preliminary estimate to choose a sample size that will give acceptably small confidence limits to the estimate of the mean.

2.2.2 Frequency of monitoring

For many purposes, an annual monitoring scheme will be sufficient; returning to the bee orchid example, there is probably no merit in undertaking two monitoring studies rather than just one each year. If, however, the monitoring is designed to detect, a posteriori, some event, such as a pollution incident in a river, then the frequency between monitoring occasions becomes of extreme importance. This is indicated in Figure 2.2. which can be considered to represent a slowly flowing lowland river. A pollution incident, such as a discharge of pesticide, will be detectable by chemical means for only a short period of time. With the river flowing, the pulse of pesticide will pass any given location rapidly and hence is unlikely to be detected by an infrequent monitoring programme. The frequency of monitoring to detect such changes in the chemical concentration of the water will need to depend upon the rate of flow of the river, but it would need to be daily or even hourly.

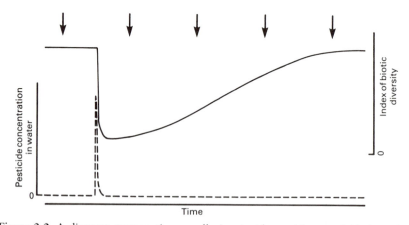

Figure 2.2 A diagram representing a pollution incident with a pesticide in a river system. Monitoring, say half-yearly, is indicated by the series of arrows. It is unlikely that such infrequent monitoring of the chemical composition of the river water (dashed line) would detect such an incident; however, monitoring of the biota (continuous line) is likely to indicate that the incident had occurred since the time-scale of its effect is much longer than the inter-monitoring period.

The pulse of pesticide will affect a range of biota, possibly eliminating some species and reducing the population density of many others. Its effects will be long-lasting as they will only disappear as the species surviving increase their population densities and as the species exterminated re-colonise the river either from upstream of the pollution incident or from neighbouring water bodies. Note that in Figure 2.2, where a twice-yearly monitoring scheme is indicated, the monitoring of biotic diversity would indicate a pollution incident, but it is unlikely to be able to indicate either when the incident occurred or how severe the incident was.

Once again there is a trade-off, this time between the frequency of monitoring (with its resource and cost implications) and the value of the data. If an incident is to be detected, then the inter-monitoring time (i.e. the time between two consecutive collections of data) should be no more than half the time that the incident's effects can be perceived.

2.3 METHOD

Most of the chapters in this book deal with the methods of monitoring. The aim of this section is to highlight three aspects of methodology that relate to the subsequent reliability and use of monitoring data.

2.3.1 Random sampling

The majority of statistics depend upon random sampling (again a full discussion of this topic is included in simple introductory statistical texts, as quoted in section 2.2.1). Figure 2.3 indicates diagrammatically the area of a nature reserve. If the aim is to record, year to year, the abundance of a plant species, say *Gentianella amarella*, in the reserve, then this can be achieved by placing a number of quadrats in which the individuals of *G. amarella* are counted. If the selection of quadrats is biased to those areas of the reserve where *G. amarella* is most common, then the estimate of abundance in the whole reserve will be an overestimate. Conversely, if the selection of quadrats is biased to areas where *G. amarella* is less common, the result will be an underestimate. What is required is the unbiased location of quadrats in the reserve.

Assuming that the intensity of monitoring allows for 15 quadrats to be recorded, Figure 2.3(a) shows a completely random location of the quadrats. Note that randomness was achieved by using a grid over the reserve, and then selecting grid coordinates from a table of random numbers. Thus, the first quadrat was selected at grid location (30, 06), where 30 relates to the horizontal axis and 06 to the vertical axis (in the illustration the origin (00, 00) is at the lower left corner). The data in Table 2.2 are based on actual counts of *G. amarella* in 0.5 m square quadrats (0.25 m^2) randomly located

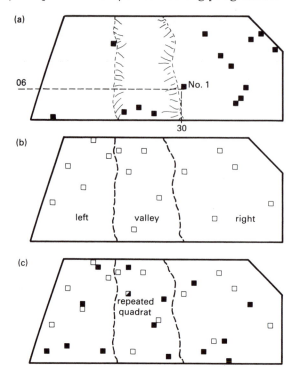

Figure 2.3 A diagrammatic nature reserve, with a valley running through it, on which a plant species (*Gentianella amarella* in the text) is to be monitored by 15 quadrats. (a) The random selection of 15 quadrats based on a square grid over the reserve; the first quadrat selected, at grid square with co-ordinates (30, 06) is indicated. (b) The reserve is now divided into three strata, with five quadrats being located at random within each stratum; this is a stratified random sample. (c) To avoid autocorrelation induced by the sampling scheme, each random selection in the three strata is used for only two successive years; open squares show quadrats recorded in years 1 and 2, filled squares quadrats recorded in years 2 and 3.

in Wharram Quarry Nature Reserve in Yorkshire. After logarithmically transforming the counts using natural logarithms, the mean number of *G. amarella* per quadrat is 2.12 which, when backtransformed, is equivalent to 8.3. When backtransformed 95% confidence limits are 5.5 (lower limit) and 12.6 (upper limit). Thus, the estimate of the abundance of *G. amarella* is 8.3 plants per 0.25 m^2, although there is only a 95% certainty that the true mean lies between 5.5 and 12.6 plants per 0.25 m^2. Much more intensive sampling would be required to narrow these confidence limits substantially. It should also be noted that the confidence interval is no longer symmetrical about the

Table 2.2 Counts of *G. amarella* in the 15 quadrats shown in Figure 2.3(a). The data have been logarithmically transformed so that they are more nearly normally distributed. Various statistical calculations are shown

Counts of G. amarella	10	8	6	2	12
(in order of collection)	8	42	7	7	24
	9	3	13	5	8
Natural logarithms of counts	2.30	2.08	1.79	0.69	2.48
	2.08	3.74	1.95	1.95	3.18
	2.20	1.10	2.56	1.61	2.08
Mean of the logarithmic counts	2.12	(backtransformed 8.3)			
Standard deviation	0.738				
Standard error	0.191				
95% confidence limits (logarithmic)	2.12 ± 0.41 (i.e. 1.71–2.53)				
95% confidence limits (backtransformed)	5.5–12.6 plants per 0.25 m^2				

mean; in the logarithmically transformed version there was symmetry (2.12 ± 0.41) but after backtransformation this property of symmetry is lost (the lower limit is 2.8 below the mean of 8.3, whilst the upper limit is 4.3 above the mean).

There is one major problem with random allocation of quadrats; sources of variation, affecting the number of *G. amarella*, may be known a priori and these are ignored by the random allocation. The diagrammatic reserve in Figure 2.3 indicates a small valley running through the centre. It might be known, a priori, that *G. amarella* is scarcer in the valley than on the higher, flatter ground surrounding the valley; should this knowledge be taken into consideration when planning the sampling programme? The random location of quadrats in Figure 2.3(a) shows many of the quadrats towards the right of the reserve (nine of the 15), with relatively few quadrats in the valley area or to the left. Intuitively, it seems possible to do better than a total random location of the quadrats!

2.3.2 Stratified random sampling

The improvement that can be introduced to the system of random sampling is to divide the reserve into strata and to sample each of these strata randomly. The predetermined strata (valley and two flat areas) are indicated in Figure 2.3(b). The 15 quadrats have been divided equally among the three strata, the locations in each stratum being selected by random allocation of grid co-ordinates. The spread of quadrats is now likely to be more uniform over the whole area of the reserve, but, as the strata have different surface areas, the statistical analysis of the data is likely to differ (see Table 2.3). The mean densities per quadrat in each of the strata can be calculated, either in the original or transformed data. These means are likely to differ, but what is important is whether the means differ significantly.

To answer such a question an analysis of variance has to be performed. With strata such as those shown in Figure 2.3(b) it is a simple one-way analysis of variance; the value of the variance ratio ($F = 1.93$; see Table 2.3) is not significant and hence the null hypothesis, that the population means of the strata are identical, is not rejected. Although apparently differing in density in the three strata, the variability of the counts (even after logarithmic transformation) within strata leads one to the conclusion that the abundance of *G. amarella* is not significantly different in the three strata. (Note that it is possible to undertake an orthogonal partition of the 'between strata' sums of squares and degrees of freedom for a comparison of valley versus top. This

Table 2.3 Counts of G. *amarella* in the 15 quadrats shown in Figure 2.3(b). Again the data have been logarithmically transformed so that they are more nearly normally distributed. An analysis of variance can be used to indicate whether the counts in the strata differ significantly

Counts of G. *amarella*					
Left stratum	8	6	12	42	7
Valley stratum	10	2	8	7	3
Right stratum	24	9	13	5	8
Natural logarithms of counts					
Left stratum	2.08	1.79	2.48	3.74	1.95
Valley stratum	2.30	0.69	2.08	1.95	1.10
Right stratum	3.18	2.20	2.56	1.61	2.08
Means (with backtransformed means)					
Left stratum	2.41	(11.1)			
Valley stratum	1.62	(5.1)			
Right stratum	2.33	(10.2)			

Analysis of Variance

Source	df	Sum of squares	Mean square	F
Between strata	2	1.8570	0.9285	1.93
Within strata	12	5.7699	0.4808	
Total	14	7.6269		

comparison has one degree of freedom, a sum of squares of 1.8402, and in comparison with the same 'within strata' mean square, $F = 3.83$. This is still not significant, though if it were slightly larger there would be evidence to suggest that *G. amarella* were less abundant in the valley than on the flat tops.) Again, introductory statistical texts will give full details about such one-way analyses of variance.

However, problems of interpretation can arise if the population is to be monitored year after year. Should one use the same set of 15 quadrats repeatedly or should new quadrats be used?

2.3.3 Autocorrelation

Take a single quadrat with approximately the mean number of *G. amarella* (say, eight) in the first year of observation. In the second year of observation the species may have become scarcer in this quadrat, say a decline of six individuals to two. What will happen in the third year? There cannot be a further decline of six; the maximum decline is two, though if the year-to-year mean density over the whole site remains constant it is possible that the number of *G. amarella* will increase. The change in the quadrat between the first and second year is, therefore, likely to exert some influence on what happens in the third year, i.e. the change between the second and third years is not completely independent of the change between the first and second years in an individual quadrat. This phenomenon is known as *autocorrelation*.

It is often preferable to plan a monitoring scheme so that such autocorrelation is avoided (such a sampling strategy for series of repeated observations is reviewed by Greig-Smith (1983)). A method of doing this is shown in Figure 2.3(c). Fifteen quadrats, five in each stratum, are randomly selected in the first year and recorded; they are recorded for a second time in the second year and that is the last time that that particular selection of quadrats is used. In the second year a new set of 15 random quadrats, five in each stratum, is selected; some of these may coincide with some of the previous quadrats if selected during the randomisation (this has happened in Figure 2.3(c) for one quadrat in the valley stratum). In the third year the second set of quadrats is re-recorded and a third set of quadrats is recorded for the first time, and so on. It can be seen that all sets of quadrats are recorded twice and twice only; hence there is no dependence of a current observation on a previous observation and no autocorrelation introduced to the dataset by the method of sampling adopted. The main drawback is that, except in the first and last years, double the intensity of monitoring is required, but this method of sampling has the advantage that the data are likely to be more reliable.

2.4 ANALYSIS

Once again, many of the chapters in this book are concerned, at least in part, with methods of analysis. It is important to realise that there are three features of monitoring data which are generally of interest, namely: **trend**, **cycles**, and **noise**. The branch of statistics known as 'time series analysis' has become extremely sophisticated with complex computer packages designed to identify and characterise these three features of time series.

However, a relatively simple pictorial example will serve to demonstrate the importance of such features. Usher (1973) used artificial moth count data to explore temporal autocorrelation; the monthly catches for a 20-year period are shown in Figure 2.4. When plotted as in Figure 2.4, the seasonality of the moth can be seen; virtually no moths in the winter months from November to March, a spring peak in May and, generally, a rather larger peak in August and September. Almost nothing can be seen about any underlying trend or about any possible cycles other than the two broods within each year; the detail in Figure 2.4 is obscuring some features of the data.

If the annual catches are plotted (see Figure 2.5), it becomes immediately obvious that there is more in the data than just seasonal variability. First, it appears that, over the 20-year monitoring period, the moth has been becoming less abundant, i.e. there is a trend towards lesser abundance. There is no simple means of saying what that trend is; it may be that there is a straight-line decrease or that there is some curved line, such as an exponential decay curve. It is up to the analyst to select a suitable model and both to fit and

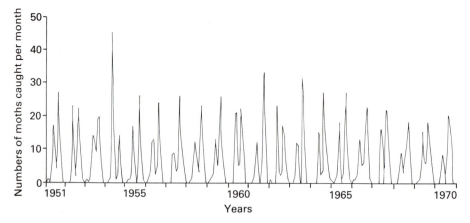

Figure 2.4 Artificial data for the counts of a moth species in a light trap. Note that the data are plotted on a monthly basis and have been collected over a 20-year period (based on Usher 1973).

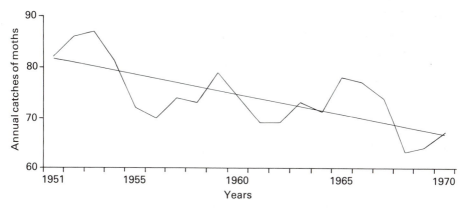

Figure 2.5 The yearly totals of the data plotted in Figure 2.4. A linear trend line has been fitted to these data (see text for the equation of this line). (Based on Usher 1973).

to test that model. The data in Figure 2.5 do not seem to warrant anything other than the simplest possible model, a linear trend of the form

$$n = a + bt$$

where n is the number of organisms (yearly total), t is the time (in years, $1951 = 0$, etc.) and a and b are constants (a positive and b negative because the number of moths appears to be ·declining). Fitting this model by a regression analysis gave

$$n = 83.4 - 0.781 \, t \, (F_{1,18} = 17.4, P < 0.001).$$

The variance ratio indicates that the null hypothesis of no linear trend can be decisively rejected. It should be emphasised that statistical significance does not imply that this linear model is the best possible model, merely that this model provides a good statistical fit to the data. Extrapolation with such a model is also unwise; the linear model above would predict that the moth became extinct 106 to 107 years after the monitoring began, thereafter having a negative population size!

Figure 2.5 also indicates that there is likely to be cyclical behaviour in this moth. Obviously there is no clearly defined cycle, since the moth is commoner than the trend predictions in years 1, 2, 3, 4, 9, 13, 15, 16, 17 and 20. An analysis of autocorrelation is required to ascertain both the periodicity of any cycles and their statistical significance.

Taking a lag of one year, a correlation coefficient can be calculated for the 19 pairs of observations of the form (year 2, year 1), (year 3, year 2), ..., (year 20, year 19). Similarly, for a lag of two years a correlation coefficient can be calculated for the 18 pairs of observations of the form (year 3, year 1), (year 4, year 2), ... (year 20, year 18). In general, for a lag of L year and with a run of 20 years' data, a correlation can be calculated between the $(20 - L)$ pairs of observations of the form (year $(L + 1)$, year 1), (year $(L + 2)$, year 2), ... (year 20, year $(20 - L)$). The series of correlation coefficients thus calculated can be plotted as in Figure 2.6; this shows clearly that there is a cycle of approximately six to seven years since there are peak correlations with lags of six, seven and 13 years.

When the trend (if any) and cycles (if any) have been eliminated, is all that remains in the data just random noise? The answer is not straightforward; the residuals should be inspected to see if there are any other systematic effects apparent. It is possible that an analysis of autocorrelation may have detected only one cycle length whereas two superimposed cycles were present. To some extent that is the case with the data shown in Figure 2.4. There is an annual cycle, though it is complex due to the bimodality of the occurrence of the moth (May and August/September generations). There is also the longer-term cycle of about 6½ years, as shown in Figure 2.6. The trend, shown in Figure 2.5, might be a real trend with the moth becoming scarcer, possibly due to habitat loss or climatic change, or it might be part of a longer-term cycle (say 50–100 years) that in a comparatively short run of data is confounded with trend.

Monitoring programmes often aim to determine trends. Analysis of monitoring data should aim to separate the three features that contribute to the value of any individual observation; the effect of trend, the effect of one or more cycles, and the residual variation that may be viewed as 'noise'.

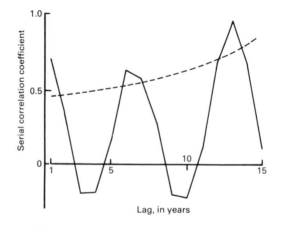

Figure 2.6 A correlogram for the data on annual counts shown in Figure 2.5. The continuous line joins the individual serial correlation coefficients and the dotted line gives an approximate limit for significance $(P = 0.05)$ of these coefficients.

2.5 INTERPRETATION

Statistical tests have been emphasised throughout the foregoing sections; significance allows a null hypothesis to be rejected, but care needs to be exercised in deciding what the alternative to that null hypothesis really is. In this section two aspects of interpretation will be considered; the first relates to finding a 'common currency' so as to be able to compare observations. The second relates to appropriate standards against which to compare observations. In both cases these can be important if a null hypothesis is to be tested and possibly rejected.

2.5.1 A 'common currency'

It is often easy to collect data that are not strictly comparable. Take, for example, the frequently used species richness, S, of an area of land. We know from many studies of island biogeography (Connor and McCoy 1979) that the number of species on an island, S, is related to the land area of that island, A, and that a similar relationship holds for many nature reserves (Usher 1985). Although there is no simple model that relates S to A, the existence of a correlation between these two variables does mean that in monitoring for species richness, similarly sized areas should be compared.

Another problem with species richness is that it does not include information on the commonness or rarity of the species. Take, for example, a woodland with 1000 trees, 995 of one species and one each of five other species. For this wood $S = 6$. In another similarly sized wood, there are approximately equal numbers of six species (i.e. each species has between 160 and 170 individuals). Again $S = 6$. Intuitively one feels that there is a difference, and hence that S is not a very good variable to use when comparing the woods.

Many possible variables that can be used in monitoring programmes suffer from similar faults, and hence attempts are made to find appropriate indices that incorporate a set of information. In the case of the two forests quoted above, a diversity index (see Magurran 1988) may be appropriate since the index will combine information not only on the number of species but also on the distribution of individuals amongst these species. It is important to find an acceptable 'common currency' to use in making any comparisons.

2.5.2 Comparison

Often statistical tests are not appropriate; monitoring continues, the volume of data grows, and it becomes increasingly obvious that something is changing (e.g. statistical significance) but one wants to know why (i.e. what is the biological significance?).

Take, for example, common bird census data (see Chapters 7 and 12) collected on many sites in the UK. An example of such monitoring is the annual study of the birds of Hopewell House Farm, a demonsration farm in the Vale of York. Figure 2.7 shows the estimated numbers of pairs of four species of birds over the monitoring period from 1979 to 1986; it appears that the skylarks have declined on this farm. It is a demonstration farm, aiming to show how wildlife and landscape values can be combined with commercial farming, and so one could ask what has been done that so drastically affects this species. Alternatively, the lapwing appears to have increased; what has improved the farm for this species? Also, both the partridge and wren seem to 'jump' around without any particular trend;

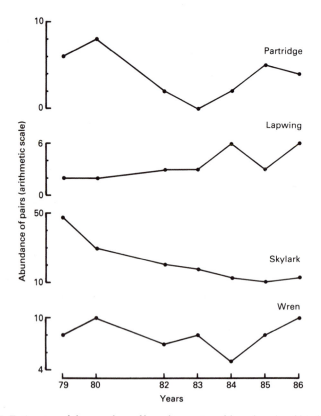

Figure 2.7 Estimates of the number of breeding pairs of four farmland bird species on the Hopewell House Demonstration Farm, Yorkshire. The estimates are based on Common Birds Census methodology (Marchant 1983), with analysis following the methods advocated by North (1977). The numbers are plotted on an arithmetic scale; there was no census in 1981.

indeed it could be that they show cycles with the partridge possibly having a longer periodicity than the wren. The only results of such questioning are speculations about what is happening on the individual demonstration farm.

If the data are put into a context, using the national Common Birds Census scheme (Marchant 1983), then interpretation of the data becomes clearer (Figure 2.8). Nationally, the skylark has declined over the period from 1979 to 1986; data from the one farm conform to the overall national position of this species. Nationally, the lapwing appears to have an upward then downward trend; the increasing trend on the farm is therefore likely to be 'real' and the reasons worth exploring. Nationally, both the partridge and wren seem to have a cyclic movement in their index value, reasonably closely correlated with the Hopewell House Farm data (especially the partridge). Again, both of these species on the farm seem to conform to the national situation.

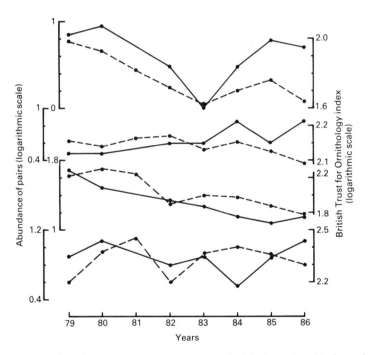

Figure 2.8 The data shown in Figure 2.7 (compared with the national index values of the Common Birds Census (dashed lines) (data from Marchant 1983 and later newsletters of the British Trust for Ornithology). The numbers of birds at Hopewell House Farm are plotted on a logarithmic scale (log ($n + 1$) is plotted because of a zero count for partridges in 1983); the BTO index values are also plotted on a logarithmic scale as advocated in Chapter 12.

There is likely to be an increasing number of national monitoring schemes, providing annual index values for birds, butterflies, etc. They are important, especially if regional index values can be derived as well, for the interpretation of local monitoring data. Geographically, Britain as a whole is probably too large for a simple national index value to be valuable (see Taylor (1986) for maps of the year-to-year patchiness in the abundance distribution of some insect species). It would therefore be preferable to compare the Hopewell House Farm data (Figures 2.7 and 2.8) with a north of England, or even a Yorkshire index, rather than a national one. Standards against which to compare individual monitoring programmes are important.

2.6 FULFILLMENT

When should a monitoring scheme be stopped? Monitoring gains a momentum of its own and it is probably true to say that 'next year's data are likely to be the most crucial of all'! However, resources are limited, it is fun to start monitoring, but when will the monitoring programme be completed? As part of the planning for any monitoring scheme, there should be rules for stopping.

The kinds of questions which need to be asked are 'How will I know when I have collected enough data?' or 'How will I know if the monitoring scheme is still relevant?' There are, therefore, two fundamentally different kinds of rules for stopping that can be built into the plans for any monitoring scheme.

First, there is the terminal kind of rule; when a criterion is satisfied then the monitoring stops. This was essentially the case for the farmland bird monitoring shown in Figure 2.7; eight years of data were considered to be sufficient for demonstration purposes, and hence the monitoring programme ended in 1986. Second, there is the review kind of rule; this avoids providing a criterion that at some stage will indicate that the decision 'stop now' is correct. Instead, it makes provision for a periodic review of the programme, at each review assessing the resources devoted to, and the benefits derived from, monitoring, and basing decisions upon some sort of a cost–benefit comparison.

A recommendation for anyone considering starting a monitoring scheme is therefore, at the same time to consider when it should be stopped. If a defined end-point can be recognised, then use that to define a terminal stopping rule. If no clearly-defined end-point can be conceived, then use a review type of stopping rule. Above all, know when sufficient data have been collected, and ensure that these data are used for their maximum value.

2.7 DISCUSSION

This chapter has been concerned with five questions that one should ask when considering any new monitoring programme (or, indeed, if one is inheriting a monitoring programme from someone else). There are, however, two further aspects of monitoring that have not been addressed since they cut across a number of these questions.

First, simple schemes are likely to be the most effective. It is, therefore, important to consider this concept and ask questions such as 'Is this the simplest method?' or 'Is this the simplest form of analysis?', etc. Simplicity will often mean ease of collection of data, ease of analysis of those data, ease of interpretation of results and, above all, ease of handing over responsibility to someone in a more subordinate position. Monitoring is usually a long-term exercise, which therefore frequently involves many different people, each possibly only for a few years.

Second, one should understand possible philosophical problems with the data. A count of *Ophrys apifera* in a quadrat or a concentration of NO_3^- in river water present no particular philosophical problems. However, some monitoring is based on recording the presence or absence of species. Presence is 'good' data, since the sight of a species means that that species was there (assuming, of course, that identification was correct). Absence is 'bad' data, since no record does not necessarily mean that the species was not there. Reliance can be judged on the size of a standard error, as discussed in the first sections of this chapter, but it is also related to the type of data being recorded periodically by the monitoring programme.

ACKNOWLEDGEMENTS

I should like to thank Dr T.J. Crawford for critically reading a draft of this chapter. The ideas in the chapter were largely developed by taking part in a monitoring course, held at University College, London in March 1989 and organised by Dr F.B. Goldsmith. I should like to thank many members of the Harrogate Naturalists' Society who undertook the CBC work on which Figures 2.7 and 2.8 are based.

REFERENCES

Bailey, N.T.J. (1981) *Statistical Methods in Biology* (2nd edn), Hodder & Stoughton, London.

Campbell, R.C. (1974) *Statistics for Biologists* (2nd edn), Cambridge University Press, Cambridge.

Connor, E.F. and McCoy, E.D. (1979) The statistics and biology of the species-area relationship, *American Naturalist*, **113**, 791–833.

Greig-Smith, P. (1983) *Quantitative Plant Ecology* (3rd edn), Blackwell, Oxford.

Magurran, A.E. (1988) *Ecological Diversity and its Measurement*, Croom Helm, London and Sidney.

Marchant, J. (1983) *BTO Common Birds Census Instructions*, British Trust for Ornithology, Tring.

North, P.M. (1977) A novel clustering method for estimating numbers of bird territories, *Journal of the Royal Statistical Society, series C (Applied Statistics)*, **26**, 149–55.

Parker, R.E. (1979) *Introductory Statistics for Biology* (2nd edn), Edward Arnold, Sevenoaks.

Taylor, L.R. (1986) Synoptic dynamics, migration and the Rothamsted Insect Survey, *Journal of Animal Ecology*, **55**, 1–38.

Usher, M.B. (1973) *Biological Management and Conservation*, Chapman and Hall, London.

Usher, M.B. (1985) Implications of species-area relationships for wildlife conservation, *Journal of Environmental Management*, **21**, 181–91.

— 3

Remote sensing techniques for monitoring land-cover

JONATHAN T. C. BUDD

3.1 INTRODUCTION TO REMOTE SENSING

3.1.1 Radiation and reflectance

Remote sensing is both an art and a science. It can be defined as a technique for obtaining data on an object without coming into direct contact with it. Sight is a form of remote sensing. Fundamental to remote sensing is the fact that in different wavelengths objects reflect or emit varying amounts of radiation. This unique pattern for a particular object is known as its spectral 'signature'.

Almost all the radiation wavebands can be used for remote sensing, ranging from radar to microwave (see Figure 3.1). Similarly, remote sensing applications are found in a wide variety of disciplines. This chapter however concentrates on the visible and infra-red wavebands, those most frequently used for ecological monitoring.

Much of the earth's surface is covered by vegetation. Vegetation absorbs, transmits and reflects solar radiation: the proportion that is reflected is determined primarily by the spectral characteristics of the vegetation canopy. It is important to remember, however, that other environmental factors also influence the amount of reflectance, in particular the amount of solar energy reaching the vegetation canopy. On a cloudy day the spectral characteristic of a canopy will be very different from that of a sunny day. Vegetation has a very characteristic spectral signature (Figure 3.2). Reflectance in the visible bands is relatively low compared with the near infra-red wavebands. The troughs in the infra-red wavebands are caused by atmospheric water absorbing a large proportion of the incoming solar radiation.

Effective use of remote sensing can only be achieved if the user understands the relationship between the characteristics of a vegetation canopy and its spectral signature as measured by the sensor.

The proportion of live and dead vegetation and bare substrate are the three most important components determining the reflectance from a vegetation

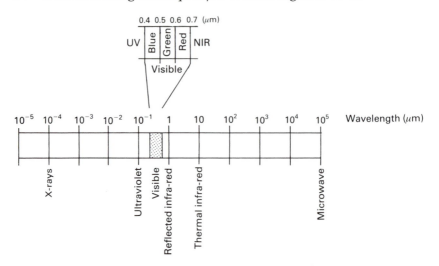

Figure 3.1 The electromagnetic spectrum – major divisions for remote sensing.

canopy (Figure 3.3). Of these, live vegetation is the most significant since it determines over 50% of the reflectance from most canopies. Reflectance from live vegetation is dependent on water content, cell structure, chlorophyll content, and plant structure (Gausman 1977).

3.1.2 Sensors and their platforms

To measure the reflectance of an object requires a sensor, which in turn must be supported on a platform. There are a wide variety of sensors and platforms to choose from. The type of sensor and the platform determine the kind of information that may be recorded.

The two most important components of a sensor are its spatial and spectral resolution. The spatial resolution is defined by the smallest feature that can be 'seen' by the sensor. The spectral resolution is the range and number of bands within the spectrum to which the sensor is sensitive. These aspects will be discussed further in the following sections.

There are also a wide variety of platforms; for example, balloons, kites, aircraft, and satellites. These platforms can be categorised according to the distance they are from the object being measured (Figure 3.4).

3.2 Satellite sensors

Scanners are most frequently used in satellites because this eliminates the problem of returning photographic film to earth. The digital data from the

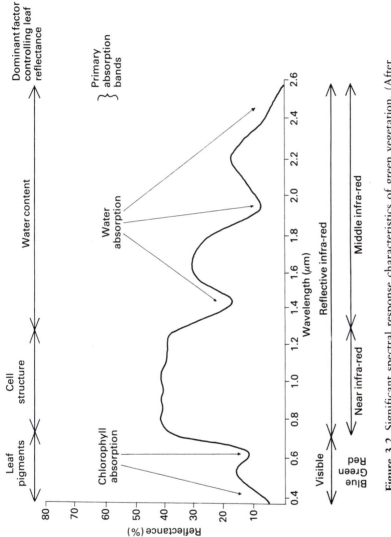

Figure 3.2 Significant spectral response characteristics of green vegetation. (After Hoffer and Johannsen 1969.)

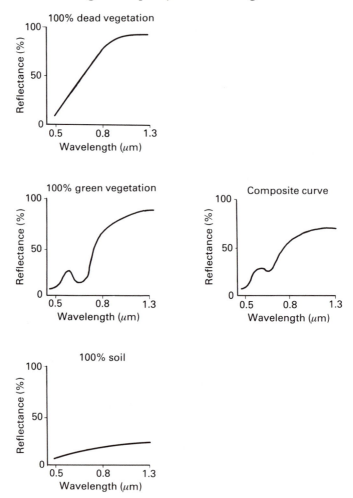

Figure 3.3 Schematic of the contribution of the individual materials to the composite spectra reflectance of the field of view of a field spectrometer. (After Pearson and Miller 1971.)

scanners can be transmitted back to earth using radio waves. Cameras have been used in manned satellites, for example space lab (Keech 1985). Only a limited amount of this type of imagery exists, however.

Figure 3.5 shows the basic structure of a scanner, and how it forms images of the earth's surface. A lens scans at right angles to the direction of movement of the platform. This lens helps focus the image on to an array of photoelectric cells which measure the amount of radiation being received.

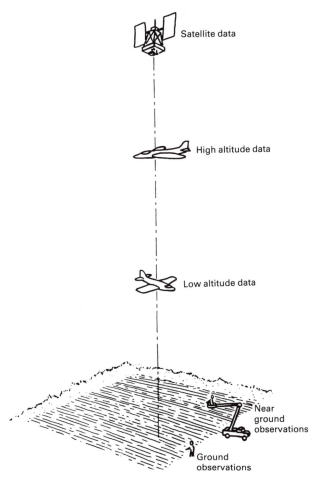

Satellite data

High altitude data

Low altitude data

Near ground observations

Ground observations

Figure 3.4 The multistage remote sensing concept. (After Lillesand and Kiefer 1979.)

This information is then converted into digital data in readiness for transmission to earth when contact is made with a receiving station.

The first satellite to be designed for environmental monitoring was Landsat 1, launched in 1973. Since then, other satellites with similar roles have been launched, including SPOT, and other satellites in the Landsat series (Table 3.1). As can be seen there has been a steady improvement in the spatial and spectral resolution of these sensors.

Many countries maintain an archive of good cloud-free images. These centres can carry out searches of their archive in order to identify suitable

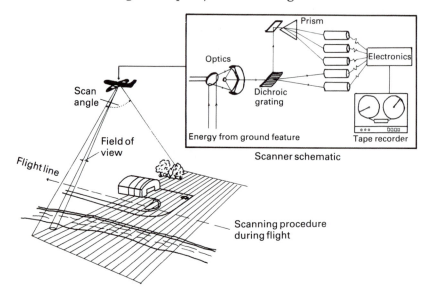

Figure 3.5 Multispectral scanner system operation. (After Lillesand and Kiefer 1979.)

imagery for a particular user's requirement. Image-processing facilities are often provided at these centres. Universities and other research establishments often have image analysis facilities and are often willing to undertake contract work. There is also a wide range of hardware and software available if a capability for this type of work is needed in-house.

The main advantage of satellite imagery is that the raw data are digital and can be analysed using computers. If the user knows that a particular area on an image is coniferous plantation then the computer can be programmed to find all other areas with similar reflectance characteristics. Alternatively the computer can be asked to identify clusters of similar reflectance throughout the image automatically (unsupervised classification). These clusters will normally represent a particular object on the ground, i.e. coniferous plantation.

One of the most obvious uses of this type of classification is for mapping vegetation. Several of the sensors have been designed specifically with that aim in mind. In large continental areas, like Africa, remote sensing has proved an ideal tool for general vegetation mapping. In Great Britain, however, satellite imagery has had a more limited success. The main reason for this has been the complex heterogeneous nature of the countryside. This was particularly a problem with early low resolution data. Even with the introduction of higher resolution sensors, there were still problems. The increased resolution led to increased variability in the reflectance. A single

Table 3.1 Remote sensing satellite sensor specifications

Satellite platform	Sensor	Spatial resolution (m)	Spectral resolution (µm)		Altitude (km)	Repeat (days)
Landsat 1–5	Multi-spectral scanner (MSS)	80 × 80	Band 4 Band 5 Band 6 Band 7	0.50–0.60 0.60–0.70 0.70–0.80 0.80–1.10	919	18
Landsat 4–5	Thematic mapper (TM)	30 × 30 120 × 120 30 × 30	Band 1 Band 2 Band 3 Band 4 Band 5 Band 6 Band 7	0.45–0.52 0.52–0.60 0.63–0.69 0.76–0.90 1.55–1.75 10.40–12.50 2.08–2.35	705	16
SPOT-1	Multi-spectral	20 × 20	Band 1 Band 2 Band 3	0.50–0.59 0.61–0.68 0.79–0.89	832	26*
	Pan-chromatic	10 × 10	Band 1	0.51–0.73		

Launch and termination dates for remote sensing satellites:

Landsat 1	23 July 1972	6 January 1978
Landsat 2	22 January 1975	25 February 1982
Landsat 3	5 March 1978	31 March 1983 (standby mode)
Landsat 4	16 July 1982	TM sensor failed February 1983
Landsat 5	1 March 1984	
SPOT-1	22 February 1986	

* Sensor can be pointed to give coverage on 11 successive days at a latitude of 45°.

field may have a wide variety of spectral patterns whereas the coarser resolution scanners tend to average out these differences.

There have been several attempts in recent years to carry out national or regional vegetation surveys with varying success. Hunting Technical Services (1986) as part of the Monitoring Landscape Change project commissioned by the Department of the Environment and the Countryside Commission evaluated Landsat TM imagery as a possible method of mapping broad land types with limited success. It was clear that certain vegetation types were far more easily identified than others. For example, it can be extremely difficult to distinguish heather moorland from coniferous plantation at certain times of the year. On the other hand it is often possible to distinguish different ages of heather (Weaver 1987). There are several reasons why it can be difficult to discriminate even the most basic vegetation categories. The reflectance from the vegetation canopy is not solely dependent on the structure of the vegetation (see previous paragraph). Other environmental components are important in determining the reflectance. The output from the sun remains fairly constant, however, the transparency of the atmosphere can vary tremendously. Even when there is no apparent cloud cover, variations in concentrations of dust or water vapour can cause significant differences in the amount of radiation reaching the earth from the sun and in turn how much is reflected back. The attenuation is not equal across the spectrum: water vapour can reduce reflectance in parts of the infra-red spectrum quite significantly. Because it is difficult to measure the opacity of the atmosphere at any one time, it is almost impossible to compensate for these effects.

The sun's angle to the earth's surface varies both seasonally and diurnally. The changing sun angle can have a profound effect on the reflectance from a vegetation canopy (Egbert and Ulaby 1972). It is extremely difficult to model these effects since they are changing constantly. Because of this it is important when measuring change from satellite images that environmental conditions prevailing at the time of data capture are similar.

Difference in the angle between the sun and the object due to the slope of the ground can also significantly affect the spectral characteristic of an object (Jones, Settle and Wyatt 1988). For example bracken on a north-facing slope will have a completely different spectral pattern to a similar canopy on a south-facing slope. This can severely limit the application of remote sensing in upland areas. Attempts have been made to reduce these effects of topography by applying mathematical models to the data, but with a limited amount of success.

The season is also important, particularly when trying to measure change. Images recorded at different times of the year can vary quite considerably. The two images shown in Figure 3.6 were recorded as part of the National Remote Sensing Centre SPOT simulation programme (Budd 1985). They show part of Skiddaw in Cumbria in May and June 1984. The upland area is to the left of each image with agricultural land appearing in red. Healthy green vegetation normally appears red on an infra-red false colour image like these. To the right of the image is a large conifer plantation. Within the upland area on the May scene heather moorland is represented as black and moorland grass as light blue. On the July scene it is also possible to see blanket mire which appears as a deep red. The light blue areas in the agricultural area on the July scene are where grass has been cut. There has been little change in the reflectance from the conifer plantation.

The above problems explain to some extent why certain habitats are more suitable for remote sensing than others. For example satellite data has proved

(a)

(b)

Figure 3.6 Simulated false colour SPOT image (bands 1, 2, and 3) of Skiddaw, Cumbria, (a) May and (b) July 1984.

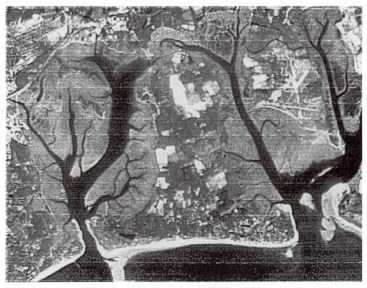

(a)

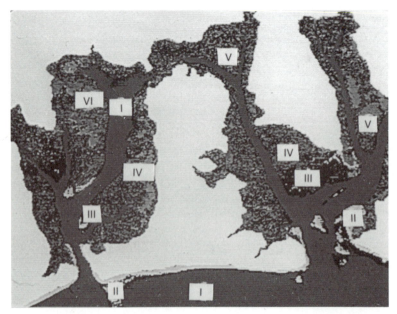

(b)

extremely useful for mapping and monitoring saltmarsh vegetation (Coulson *et al.* 1980). The fact that there is little topography, and relatively homogeneous mono specific vegetation types, has particularly suited remote sensing. It is possible to map effectively some vegetation types with a high degree of accuracy. Figure 3.7 shows a Landsat TM false colour image of Langstone and Chichester Harbours in southern England. This image has been classified into six cover classes.

3.3 AERIAL SENSORS

Air photo cameras are the most frequently used sensors in aircraft (Figure 3.8). As with the scanners, lenses focus the image on to the recording medium, in this case photographic film. The spectral range of photographic film is limited, compared with that of the electronic scanners. Most films are only sensitive to the visible wavebands though some specialist films are sensitive to near infra-red light. There are three main types of film; black and white panchromatic, normal colour, and infra-red false colour. Black and white photography has been available since the later part of the 19th century. The two-colour types of film are more recent.

Resolution of the film is far greater than the achievable using scanners (see section 3.2). Black and white aerial photography at a scale of 1 : 20,000 has a spatial resolution of at least 25 cm on the ground. Colour films have a slightly lower spatial resolution of 30 cm at the same scale. Colour films also require better atmospheric conditions in particular low levels of haze.

The two most frequently used formats for aerial photography are 9 × 9 in (135 mm × 135 mm) and 70 mm. The first is best suited to covering large areas and providing a high degree of spatial accuracy. The smaller format is cheaper to operate and can be successfully used for more localised monitoring.

James Wallace Black was one of the first to use aerial photography when, in 1860, he obtained photographs of Boston from a balloon (Newhall 1969). It was, however, military use during World Wars I and II that led to the greatest progress in the acquisition and interpretation of aerial photography. It was in the immediate post war period that the RAF obtained comprehensive air photo cover of Great Britain. The Ordnance Survey flew over most of Great Britain again as part of their mapping programme in the late 1960s and

---

Figure 3.7 (a) False colour Landsat TM image (bands 4, 5, and 6) of Langstone and Chichester Harbours.
(b) Classification of the above image. I = water, II = sand and gravel, III = mud, IV = *Enteromorpha* species and *Ulva*, V = *Spartina* species, and VI = upper salt marsh species.

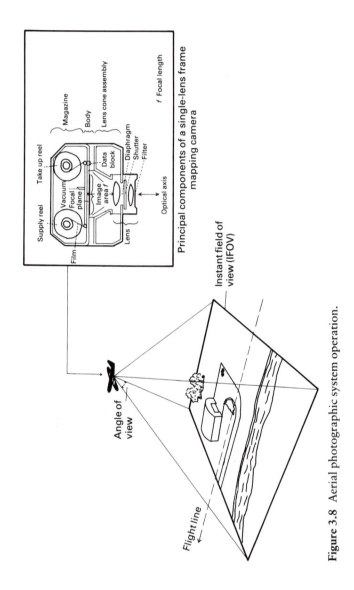

Figure 3.8 Aerial photographic system operation.

throughout the 1970s. In the late 1980s there have been a wide variety of flying programmes. Many counties and regions have been flown over as have National Parks and Environmentally Sensitive Areas (ESAs). In 1988–9 complete coverage of Scotland was obtained. Almost all the above aerial photography has been black and white panchromatic. As the relative cost of colour film reduces, more surveys are using this kind of film. There are several national centres where records of air photo coverage are kept. These centres will carry out a search to identify photography of a particular area selected by the user. They will in most cases also supply the photography.

One of the greatest advantages to monitoring using aerial photography is the fact that comprehensive data exist for most of the UK since the 1940s. This provides a unique database for monitoring land cover change. A good example of how this data set can be used is provided by the National Countryside Monitoring Scheme. This project was set up by the Nature Conservancy Council in 1983 to measure the loss of wildlife habitats in the wider countryside since the 1940s. Aerial photographs were the only consistent source of data available for the 1940s and the present day. Some time was spent evaluating the level to which habitats could be interpreted from aerial photographs. The list of features in Table 3.2 is based on those features that could be identified with a reasonable degree of accuracy and are important in terms of habitat loss. Some of the features can be identified more reliably than others so it is important that the user understands any limitations of the interpretation. The hierarchical classification allows classes to be amalgamated if there are problems in discriminating between them. For example, it can be difficult to distinguish old broadleaved plantation from semi-natural broadleaved woodland. The hierachical structure allows reliable estimates to be given for the combined class of broadleaved woodland, whatever its origin.

The project uses a sampling approach which is the most efficient way of obtaining the data. A census would not have been practical because of the costs involved. Even with a 10% sample as used by this project the task is considerable. In order to reduce sample variance the sample is stratified into broad land types. Satellite imagery was classified to give broad land types within a region or county. The sample is then selected strictly randomly from within each of these strata. Typical land types are moorland, intensive agriculture, forestry, and urban. The latter is particularly important in that it is around urban areas where the greatest change often takes place.

Once the sample has been selected searches were made of the air photo archives to find the necessary photography. Ideally there should be complete air photo cover for the area under study from which to select the sample. If this is not the case then bias may be introduced into the sample. For example, the Ordnance Survey fly aerial photography in order to update their maps and so for this reason there is a higher proportion of change on these

Table 3.2 List of feature types to be recorded

Group A	Group B
Linear features	
Hedgerow	Hedgerow without trees
	Treeline, including hedgerow with trees
Water	Running natural water
	Running canalized water
Unsurfaced tracks	
Area features	
Woodland	Semi-natural broadleaved woodland
	Broadleaved plantation
	Semi-natural coniferous woodland
	Coniferous plantation
	Mixed woodland
	Young plantation
	Recently felled woodland
Parkland	
Scrub	Scrub, tall
	Scrub, low
Bracken	
Heathland	Heather moorland
	Montane heath
	Maritime heath
Mire	Blanket mire
	Lowland raised mire
Wet ground	
Marginal inundation	
Open water	Standing natural water
	Standing man-made water
	Running natural water
	Running canalized water
Grassland	Unimproved grassland
	Semi-improved grassland
	Improved grassland
Arable	
Bare rock and soil	Unquarried inland cliff and rock outcrop
	Quarries and open-cast, including spoil
	Other bare ground
Built land	Built land
	Recreation
	Transport corridors

photographs. The stratification does help to reduce this problem by ensuring that the estimates are weighted in proportion to the area covered by that land type. So far there has been good air photo cover for all the areas studied.

Interpretation of the photographs is done in two ways. In flat lowland areas which have significant amounts of detail on the OS 1 : 10,000 series maps, the information is transferred directly from the aerial photographs on to the maps using field boundaries, etc. to ensure the correct location of the parcels of land. In upland areas, or areas devoid of topographical detail, then another approach is required. Photogrammetric plotting machines (Lo 1976) are used to interpret and map the habitats. These machines automatically correct for distortions inherent in the aerial photographs. They also correct for scale changes due to topographic effects. Because the machines can accommodate these distortions they can be used to precisely map boundaries, corrected to OS national grid, thus eliminating the need to use map detail to locate the vegetation boundaries. The reason why this machine is not used for all mapping is that it takes roughly twice as long to do the interpretation because of the time needed to set the machine up in readiness for interpreting the photographs.

Air photo interpretation is a skilled job and all staff go through a long training programme. Before work begins on any region or county the team involved spend time in the field familiarising themselves with the type of countryside that they will be working in. The operators visit the field again when they have completed the initial interpretation of the aerial photographs. Each individual sample square is visited in order to check areas of doubt in the interpretation.

One of the biggest problems encountered in previous similar surveys (Langdale-Brown *et al.* 1980) was that of accuracy of area and change measurement. For this reason the NCMS uses a computer-based digital mapping system to calculate the areas of habitats and measure the changes that have taken place. Digitising tablets are used to convert the analogue habitat maps into a digital form. The data are stored within the system as polygons (areas) and lines which are encoded with the identity of the habitat that they represent. It is important to ensure that these polygons are closed before any area measurements can be made. Each polygon or line is encoded during data capture. These data are then stored within the personal computer for analysis. The software is used to check the data for errors which are then highlighted and the operator is asked to correct them. Once the data are correct the computer can be used to calculate the area or length of each habitat. A second programme is used to check that all the polygons have been encoded or that they do not contain more than one code. Once the data sets for the 1940s and the 1970s have been verified then the computer is used to measure the change that has occurred. It is at this point that the need for

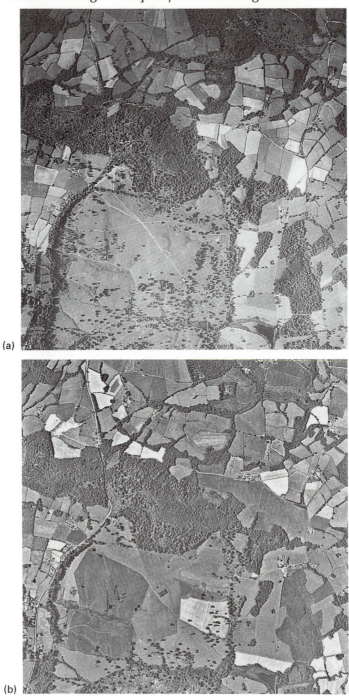

(a)

(b)

precise mapping of boundaries becomes apparent. If a boundary has an error displacement of 1m, this could cause a change to be recorded. The problem is particularly acute with linear features. Considerable effort has to be expended in ensuring that boundaries that have not changed have identical locations in both time periods. In upland areas, where boundaries are often transitions, there may be a discrepancy in the location of the boundary from one time period to another.

The result is a matrix of change for both the linear and area features. The matrix for each sample square is then used to calculate the overall estimates for the county or region. In simple terms, the estimates are calculated by weighting them in accordance with the area covered by each stratum. Standard errors are also calculated to give an estimate of sample variance (Cochran 1977).

Results are now available for several counties in England and for most of Scotland. It has long been apparent that hedgerows are being removed and

(c)

Figure 3.9 Part of the air photo cover for a sample square in East Sussex. The three photographs were flown in (a) 1945, (b) 1972 and (c) 1985. (Reproduced from an Ordnance Survey aerial photograph with the permission of the Controller of Her Majesty's Stationery Office © Crown Copyright.)

that large tracts of land are being planted with conifers. This project has been able to quantify these changes for the first time on a county/regional basis. The results have shown not only the changes that might have been expected but has also highlighted other changes that were not so obvious.

The most stunning evidence is the proportion of land that has shown change, in many regions and counties less than 60% of the land cover has remained unchanged. In some counties such as Bedfordshire, there has been little post-war loss of semi-natural habitats since most had already gone by the 1940s.

In lowland areas, the greatest losses have been to agricultural land. Lowland heath has been lost almost completely from many areas, in Cumbria and Grampian for example (NCC 1987; NCC and CCS 1988). In Sussex, large areas of broadleaved woodland have been lost to agriculture ($43 \, km^2$). The three photographs in Figure 3.9 cover sample square 11 in East Sussex for 1945, 1972 and 1985. These photographs show some of the typical changes that have taken place since the 1940s. Large areas of lowland heath and broadleaved woodland have been lost to coniferous plantation and agriculture. There have been large losses of hedgerow in all lowland areas. The percentage loss has varied from -45% in the Borders region to -20% in Sussex.

Not surprisingly, built land has accounted for losses in a wide variety of countryside features. Agricultural land has suffered the most. Lothian for example has seen an increase of $83 \, km^2$ in built land of which 74% has been at the expense of arable and improved grassland. Even though transport corridors are normally narrow features they account for large losses. In the Borders region, transport corridors accounted for the loss of $15 \, km^2$ of grassland and arable.

The fringe between lowland and upland is the area that has suffered the most due to afforestation. Increases in conifer plantations have been large. Coniferous plantation in Dumfries and Galloway has increased from $55 \, km^2$ in the 1940s to $1069 \, km^2$ in the 1970s. Most of the gains in this region have been at the expense of unimproved grassland. This contrasts with Grampian region where heather moorland has been lost to conifer plantation. Conifer plantations are one of the most frequent causes of heather loss. Unimproved grassland also account for a large proportion, in Cumbria for example $209 \, km^2$ of unimproved grassland has been gained from heather moorland.

3.4 GROUND SURVEY

The great advantage of satellite and airborne remote sensing is that large areas can be covered at any one time which contrast with conventional ground survey using ecological field methods. What, it may be asked, then, is

Figure 3.10 The radiometer in use over a stand of *Spartina* × *townsendii* agg.

the use of ground-based remote sensing (i.e. cameras and radiometers)? The answer is that there are two main uses, firstly to provide ground truth data for satellite and airborne remote sensing and secondly to make nondestructive measurements of plant characteristics, in particular biomass.

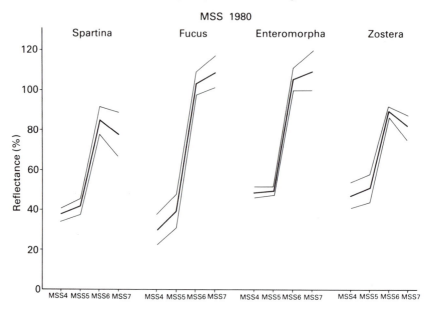

Figure 3.11 Univariate plots of mean reflectance +/− 1st standard deviation for each species, data collected August 1980.

Ground-based photography is one of the simplest forms of remote sensing available to the environmental scientist. Oblique or vertical photographs can provide a record of vegetation cover, structure, and vigour. Vertical ground photography is particularly suited to ground truthing for aerial photography. For example, infra-red photography has been used to determine the best season for discriminating salt marsh vegetation from aerial photography. It was possible to determine the emergence of certain species, i.e. *Zostera marina L.*, and also the season of greatest contrast between species. One problem with ground photography is its high spatial resolution which is not representative of the resolution of aerial photographs or more particularly satellite data. For example with ground photography it would be possible to see the spectral characteristics of each individual component of the canopy (leaves, stems etc). On a satellite image these would be combined to give an average spectral response. Also ground photography is limited by the spectral sensitivity of photographic film as with aerial photography. Both these factors limit its use for the ground truthing of satellite images.

The alternative is ground radiometry (Figure 3.10), which has two advantages over ground photography. Firstly, quantitative measurements of reflectance are obtained and secondly these values represent the average

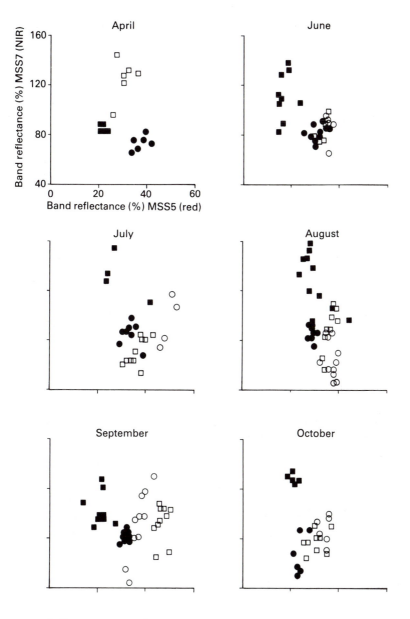

Figure 3.12 Bivariate plots of infra-red against red reflectance for four saltmarsh species during 1981.

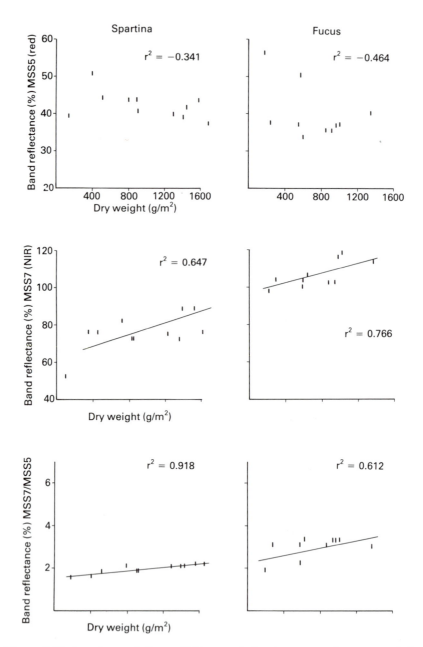

Figure 3.13 *Spartina* and *Fucus* 1980, correlation between reflectance and dry weight.

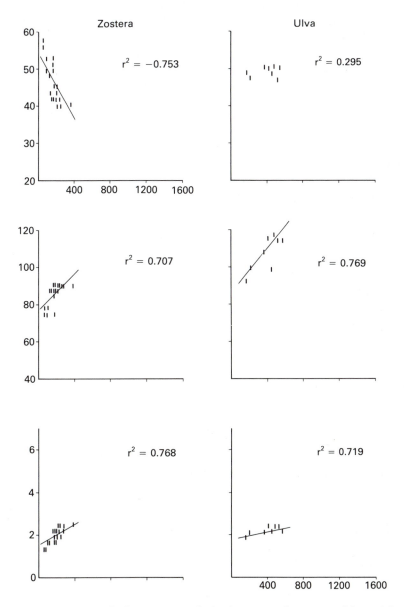

Figure 3.14 *Zostera* and *Ulva* 1980, correlation between reflectance and dry weight.

reflectance over the total field of view of the sensor. This means it is possible to simulate the individual pixels (picture cells) of a scanner. Figure 3.11 shows the mean spectral reflectance for four saltmarsh species in August 1980. As with ground photography, it is possible to identify the best season for discriminating between vegetation types. In Figure 3.12 scatter plots of infra-red reflectance against red reflectance are shown for different times of the year. It can be seen that September is one of the better months for discriminating between the four saltmarsh species, *Spartina*, *Fucus*, *Enteromorpha* and *Ulva*. In addition, it is possible to answer questions on the effects of different environmental components on the spectral characteristics of the vegetation canopy. For example how does the solar angle affect reflectance? Visible reflectance tends to increase with solar angle, while infra-red reflectance decreases. As part of this study other factors were also examined, for example tidal exposure. It is also possible to investigate further how the different components of the canopy contribute to the overall spectral characteristics of that vegetation type. Comparisons between reflectance and certain vegetation characteristics show interesting correlations (Figures 3.13 and 3.14). There tends to be a negative correlation between visible reflectance and vegetation amount, while infra-red reflectance shows a strong positive correlation (Budd and Milton 1982). For some species the correlation is sufficiently significant for the technique to be used for non-destructive biomass estimation.

3.5 CONCLUSION

Though ground radiometry is very important for understanding the factors that control the reflective properties of an object, it has limited use for mapping large areas. The two main contenders for ecological monitoring are thus satellite borne sensors and airborne cameras. The two approaches have different advantages, which are summarised below.

3.5.1 Advantages of aerial photography

(a) *High spatial resolution*
Resolution is the smallest object visible to the sensor. For aerial photography at 1 : 10,000 scale this would be 15.9 cm compared with 10 m of SPOT, though this can vary depending on the spectral characteristics of an object. A road, for example, has a high reflectance and will influence a pixel even though only a small proportion of the pixel is road. The opposite would be true of a low reflectance object.

(b) High geometric accuracy

The largest practical scale for aerial photography is 1 : 3,000. At this scale it is possible to measure objects on the ground to +/− 4.8 cm compared with the highest resolution satellite imagery of about 10 m.

(c) Good stereo images

Good stereo images though SPOT can now produce stereo images. Possible to measure height accurately.

(d) Low analysis costs

The minimum requirement is a hand stereoscope, though more expensive equipment can be used. Also with the increase in resolution of satellite imagery it can be interpreted more like aerial photography.

(e) Historic data

There is comprehensive cover from the 1940s onwards for Great Britain but limited earlier cover. The earliest environmental satellite (Landsat 1) was launched in 1973.

3.5.2 Advantages of satellite borne scanners

1. Wide variety of spectral bands.
2. Low costs
     Details are given in Table 3.3.
3. Frequent data collection on a regular basis
     SPOT can be redirected to give data collection on a daily basis for 11 successive days. Temporal.

Table 3.3 Specifications and costs for remote sensing data

	Area covered (km²)	Unit cost (£)	Cost per unit area (pence/km²)
Aerial photography			
1 : 5,000 scale	0.42	10.00	23.80
1 : 10,000 scale	1.77	10.00	5.65
1 : 20,000 scale	6.69	10.00	1.49
Satellite imagery			
Landsat MSS (4 bands)	34,225.00	277.73	0.01
Landsat TM (7 bands)	34,225.00	1,573.78	0.06
SPOT (XS or PAN)	3,600.00	806.15	0.22

4. Digital data for processing

Allowing quantitative analysis of the reflectance values (though it is possible to scan aerial photos).

5. Integration with GIS

This can be achieved with photogrammetric equipment though this is expensive. Integrate directly with GIS without expensive data capture system, i.e. photogrammetric equipment.

Generally speaking, aerial photography is most suitable for the detailed site-specific or sample based surveys, while satellite imagery can be used for broad land use classifications or for surveys that require specific attributes of satellite imagery, for example the thermal band of Landsat TM.

With either satellite imagery or aerial photography, it is important to remember that the data only represents a snapshot in time. Under certain conditions an aerial photograph of a site may bear no resemblance to another photograph of the same site taken only a few minutes earlier. It is particularly important to bear this point in mind when using remote sensing for change measurement. Season, atmospheric effects, and sensor calibration can cause differences between images that far exceed any real changes that have taken place.

The ITE publication *Ecological Mapping from Ground, Air, and Space* (Fuller 1983) provides further examples of the applications of remote sensing. No one remote sensing technique is better than another for all ecological monitoring tasks (Budd 1987). It must be left to the user to decide which method is best. Hopefully this chapter has provided a basis from which the ecologist can begin to make this kind of decision.

REFERENCES

Budd, J.T.C. and Milton, E.J. (1982) Remote sensing of salt marsh vegetation in the first four proposed Thematic Mapper bands, *International Journal of Remote Sensing*, 3(2), 147–61.

Budd, J.T.C. (1985) SPOT simulation study for Skiddaw, Cumbria. *SPOT simulation report*. National Remote Sensing Centre, Farnborough.

Budd, J.T.C. (1987) Remote sensing applied to the work of the NCC in upland areas, in *Ecology and Management of Upland Habitats: the Role of Remote Sensing, Remote Sensing Special Publication No. 2*. Department of Geography, University of Aberdeen, Aberdeen.

Cochran, W.G. (1977) *Sampling techniques*. (3rd ed.) Wiley, Chichester. Chapter 6.

Coulson, M.G., Budd, J.T.C., Withers, R.G. and Nicholls, D.N. (1980) Remote sensing and field sampling of mudflat organisms in Langstone and Chichester Harbours, southern England, in *Systematics Association Special Volume No. 17(a), The Shore Environment, Vol 1: Methods*, (eds J.H. Price, D.E.G. Irvine and W.F. Farnham), Academic Press, London, pp. 241–63.

Egbert, D.D. and Ulaby, F.T. (1972) Effects of angles on reflectivity, *Photogrammetric Engineering*, 38, 556–64.

Fuller, R.M. (ed.) (1983) *Ecological Mapping from Ground, Air and Space. Institute of Terrestrial Ecology Symposium No. 10.* ITE, Swindon.

Gausman, H.W. (1977) Reflectance of leaf components, *Remote sensing of Environment*, 6, 1–9.

Hoffer, R.M. and Johannsen, C.J. (1968) Ecological potential in spectral signature analysis, in *Remote Sensing in Ecology*, University of Georgia Press, Athens, Georgia.

HTS (Hunting Technical Services Ltd) (1986) *Monitoring Landscape Change* (10 vols.). Final Report to the Department of the Environment.

Jones, A.R., Settle, J.J. and Wyatt, B.K. (1988) Use of digital terrain data in the interpretation of SPOT-HRV multispectral imagery. *International Journal of Remote Sensing*, 9, 669–82.

Keech, M.A. (1985) The success of the metric camera as a data source for land resource evaluation. *Photogrammetric Record*, 11(66), 721–7.

Langdale-Brown, I., Jennings, S., Crawford, C.L., Jolly, C.M. and Muscott, J. (1980) *Lowland agricultural habitats (Scotland): air-photo analysis of change.* Nature Conservancy Council, Peterborough. (CST report no. 332.)

Lillesand, T.M. and Kiefer, R.W. (1979) *Remote Sensing and Image Interpretation*, Wiley, Chichester, p. 612.

Lo, C.P. (1976) *Geographical applications of aerial photography.* David and Charles, London.

Nature Conservancy Council (1987) *Changes in the Cumbrian countryside. First report of the National Countryside Monitoring Scheme.* Nature Conservancy Council (Research and survey in nature conservation No. 6), 39.

Nature Conservancy Council and Countryside Commission for Scotland (1988) *National Countryside Monitoring Scheme, Scotland: Grampian.* Countryside Commission for Scotland and Nature Conservancy Council, Perth.

Newhall, B. (1969) *Airborne camera.* Hastings House, New York.

Pearson, R.L. and Miller, L.D. (1971) A field light quality laboratory: Initial experiment: The measurement of percent of functioning vegetation in grassland areas by remote sensing methodology, in *Colorado State University, IBP Grassland Biome, Technical Report 89*, Fort Collins, Colorado, p. 19.

Weaver, R.E. (1987) Spectral separation of moorland vegetation in airborne TM data, *International Journal of Remote Sensing*, 8, 43–55.

4

Monitoring plant populations: census as an aid to conservation

MICHAEL J. HUTCHINGS

4.1 INTRODUCTION

A recurrent and consistent theme throughout this book is that an essential component of monitoring is the repeated collection of data over time. However, there has recently been some difference of opinion in the plant ecology literature about which techniques of data collection qualify as monitoring. While there is widespread agreement that the method giving the most useful information is census, Palmer's (1987) recent review of research projects also included inventories and surveys as monitoring techniques. Pavlik and Barbour (1988) limit use of the term 'monitoring' to studies involving censuses, a view with which I agree. In this contribution, therefore, as in several other papers concerned with monitoring (e.g. Davy and Jefferies 1981; Williams 1981; Bradshaw 1981; White and Bratton 1981; Hutchings 1989a, 1990), only data collected from censuses are considered.

Pavlik and Barbour (1988) describe inventory as 'geographically based assessment of entire taxa', and survey as 'ecologically based assessment of entire taxa'. The distinctive feature of census, which neither inventory nor survey possess, even though they may involve collection of data on more than one occasion, is repeated records of the presence and performance of individual plants. From the data collected by censusing, the skilled analyst can identify the times during the life cycle when a species is performing poorly (or too successfully), and perhaps prescribe changes in management to alleviate the problem. Inventories and surveys, on the other hand, can only show what is at a site, or, if repeated records are made, what has changed from one date to another; neither provides a direct basis for understanding *what* is happening to a population.

4.2 THE NEED FOR MONITORING PLANT SPECIES

Palmer (1987) claimed an upturn in the number of monitoring projects initiated in recent years, although 60% of the projects she included in her review were inventories and surveys, rather than monitoring projects, using the term as it is defined in this paper and in Pavlik and Barbour (1988). Even so, there has been a slow increase in the number of studies involving censuses of plants. This should be set in context, in order to guard against complacency; Perring and Farrell (1983) commented that of 317 species listed in the *British Red Data Book* for higher plants, only 6% had been the subjects of accounts in the Biological Flora of the British Isles (published in the *Journal of Ecology*), and very few of these species had been the subjects of detailed autecological study. The situation has improved little since then, although some further studies have been published (e.g. Bullard, Shearer, Day and Crawford 1987; Hutchings 1987a,b). Although figures are very hard to come by, it seems likely that other temperate regions probably have a similar low level of knowledge which can be used as the basis for devising management plans for their endangered flora. The record in other latitudes, however, is certainly vastly inferior; in many tropical and polar latitudes, the preliminary tasks of inventory and survey of the species present, and the extent and location of their populations, is far from complete.

We tend to regard rare species as the most deserving of the labour and time necessary to conduct an adequate monitoring programme. The reason is clearly that we need to know how they are performing through time (whether they are declining, stable or increasing) and how to manage them sensitively for their conservation. The examples of monitoring projects quoted in this paper are all for rare species, although it should be remembered that many different criteria are used to judge whether species should be regarded as rare (Ayensu 1981; Harper 1981a; Rabinowitz 1981). Both Harper and Rabinowitz have discussed the various qualities of distribution and abundance which result in different types of rarity. It may be important, where time and funding are limited, to prioritise different types of rarity, and then monitor those species with the types which are associated with the greatest risk of extinction. It may also be wasteful of resources to monitor species which are rare on a national or county level when these species are abundant in other counties or other national territories. There are, however, species other than rare ones which are candidates for monitoring. These include species which are not at present under threat, but which have rapidly declining abundance (including some now uncommon weed species), and species with increasing abundance, such as certain aliens.

4.3   METHODS

Census methods for studying the population biology of plants were popularised, although far from pioneered (see White 1985a,b), by John L. Harper. They are far more time-consuming and labour-intensive than either inventories or surveys, and data are collected over a longer period. A decision to undertake a census study of a plant species or population necessitates a commitment to a very large amount of work, and therefore planning should be rigorous, in order to achieve the maximum in results for the effort entailed. The checklists produced by Jeffers (1978, 1979) are helpful in this respect. Palmer (1987) also offers some advice, summarized in her Table 4.2, on planning such a project, on sampling and experimental design and on the dissemination of results. (Items 4–6 of her table concern the use of inventory and survey methods, and are therefore less helpful.)

The most valuable property of census data is that they can be used to calculate demographic parameters, i.e. the statistics of births, deaths, fecundity, diseases etc. The flux of organisms through the population can be analysed together with its age structure, age-specific mortality and fecundity (Hutchings 1986, pp. 415–23). The information generated by calculating demographic parameters is indispensable for understanding why species are threatened by local extinction or why they become problematical, as well as for developing management plans to alleviate the problem. In addition, these parameters can be used to judge management success, by comparing the demography of control populations and populations subjected to altered management (White and Bratton 1981). The data can also be used to answer questions such as whether the death-rate of plants in the population exceeds the birth-rate, whether the age structure of the population indicates that recruitment of new plants is taking place, and whether seed production is limited by seed predators. Information about these and other aspects of plant population biology is important for formulating management plans, and such questions should be posed at the planning stage of the project, rather than when the results are analysed (Prince 1986; Palmer 1987).

Although rarity is often a reason for monitoring a species, it is imperative that, whenever possible, an adequate number of plants is recorded. If possible, records of presence and absence should be supplemented by measurements of performance, such as height or number of seed capsules produced. Plants are highly plastic organisms (i.e. their form is strongly influenced by their growing conditions), so that such measures are important indicators of the condition of populations. For example, Mehrhoff (1989) has shown that in the rare orchid *Isotria medeoloides*, stable populations contained a high proportion of flowering plants but declining populations contained few, and that size of an orchid in one year was a good predictor of its behaviour (flowering, arrested flowering, sterile) in the following year. Population performance measurements are often expressed as mean values, so that, when records are not made on the whole population, enough plants must be sampled to allow an accurate estimation of the mean; as variability of individual performance increases, so does the minimum acceptable sample size. It should also be remembered that the calculation and comparison of many statistics is invalidated if individual values in a sample are not normally distributed about the mean, and that in such cases transformations of the data are often necessary before statistical techniques are applied to the data. Sampling should be undertaken in a manner which is free of bias. For example, measures of size taken only from flowering plants overestimate the mean, because large plants have a higher flowering probability than small plants. More detailed advice on sampling is readily available in elementary texts on plant ecology, such as Greig-Smith (1983), Kershaw and Looney (1985), and Moore and Chapman (1986).

Detailed descriptions of the techniques relevant to censusing plant populations are provided by Hutchings (1986, pp. 395–403). This reference covers the subjects of establishing and relocating permanent plots, the tagging and mapping of plants, and the measurements which can be made on living plants without impairing their performance or survival. Different species and different circumstances may require flexibility in applying these methods or the invention of new, more appropriate ones, so that this source should not be regarded as exhaustive.

4.3.1 Problems in censusing plant populations

There are several important points to be made concerning how and what to record, and some of the pitfalls which must be avoided, when a monitoring census is undertaken.

1. Flowering individuals are only the most noticeable components of plant populations. Non-flowering plants must be located and recorded, even when this presents difficulties. There is an important practical aspect to recording these plants, apart from the fact that ignoring them will result in underestimation of population size. Non-flowering plants are usually the youngest or smallest members of the population, and thus they may be important indicators of the success of recent recruitment, both to the growing and the reproducing components of the population. In the absence of data on non-flowering plants, the incorrect conclusion would be drawn that all plants achieve sexual maturity and therefore the population would not benefit from management intended to improve this aspect of performance. If data are collected showing that a very low proportion of plants in the population are flowering, a very different form of management might be prescribed.

2. Growing plants are only the most apparent parts of many plant populations; in many cases they are outnumbered severalfold by dormant, viable seeds in the soil (Roberts 1981). Hutchings (1986 pp. 378–95) and Thompson (1986) discuss sampling and recording the size of this component of plant populations, although to date, methods for censusing individual seeds in small-seeded species have not been devised. However,

the rates of loss of seeds from samples sown into sites, or cohorts of seeds all produced during specified time intervals (e.g. one growing season) have been calculated by spraying the seeds with paint or dye, or labelling with radioactive markers (Watkinson 1978a,b). In some species there are also dormant bud and meristem banks which provide the potential for population increase.

3. Flowering takes place in many species at the end of a period of above-ground growth. Thus, plants which still possess aerial parts during the flowering season may only be a small proportion of those present at the start of the period. There may either be considerable mortality between commencement of above-ground growth and flowering, or, in perennating species, underground parts may survive although all aerial parts have been lost. Where mortality has occurred, collecting data only during the flowering season will not reveal the important periods of mortality, or enable their causes to be determined, possibly preventing effective management plans from being devised. The solution is to repeat censuses at regular intervals during the growing season, as in the exemplary study by Leverich and Levin (1979) on the annual *Phlox drummondii*. Developmental stages are tightly compressed in time in annual species, and it may be wise to conduct censuses as frequently as once per week. In some perennials, particularly those with persistent aerial structures, it may be sufficient to census populations once a year (or even less regularly in trees and shrubs). However, in herbaceous perennials which lose all aerial parts but perennate below ground, censusing only during the flowering season can seriously underestimate the size of the surviving population (Hutchings 1987a). Thus, although we often collect data only during flowering periods, largely because it is instinctive to record what is attractive, this is often not a sound way to census a population of plants.

4. It is difficult in many species to accurately assess the flux of seedlings in populations, because the life spans of so many recruited seedlings are extremely short. In many cases seedlings appear to be eaten by herbivores, infected by pathogenic fungi or desiccated within hours of germination. A complete census of these very short-lived seedlings would necessitate the undertaking of almost continuous recording. In most cases this will clearly be impractical, and a compromise must be reached between accurate assessment of seedling flux and practicalities. The major impact of these short-lived seedlings upon a plant population is to deplete the seed bank while not contributing to population replacement.

5. Some perennial species, notably orchids, can enter lengthy periods of undergound dormancy within their life spans, during which they persist only as underground storage organs. During such periods, which may last for one or more growing seasons, census does not reveal these plants,

leading to the suspicion of their deaths. At the end of the dormant period, however, they can reappear, forming aerial parts once more (Tamm 1972; Wells 1981; Hutchings 1987a). Fortunately, these dormant periods rarely last for more than two years, so that plants which do not appear above ground for three consecutive years can be regarded as almost certainly dead. However, the problem with such dormancy is that, until many years of census data have been accumulated, there is uncertainty about when plants were recruited, because germination early in the study cannot be distinguished from emergence from dormancy. Similarly, towards the end of a run of census data, it is not possible to determine whether plants absent during recent censuses are dead or dormant.

6. A final problem, once again common amongst orchids, is lengthy periods between germination and the date of first production of aerial structures. Wells (1981) has tabulated results showing that in some species of orchid a decade may pass between germination and the production of the first leaf, although for others the period may only be one or two years. Again, where this type of behaviour occurs, it creates problems in analysing population flux, as well as in calculating the age structure of the population. It also causes difficulties in assessing the success of a particular form of management, since changes in recruitment rates may not be visible for several years after the management is introduced.

4.3.2 Units to record in plant populations

It is possible (e.g. De Steven 1989) to distinguish between censuses and demographic analyses undertaken at the level of the genet (the genetical individual, represented by all tissue ultimately derived from a single seed), and at the level of the ramet (the repeated structural units from which plants capable of vegetative propagation are constructed). In plant species without vegetative propagation, the units to be recorded in a census are straightforward. Each unit is a genet, and there are obvious distinctions between genets. In species which grow clonally by vegetative propagation, the genet may consist of any number of ramets and the distinctions between genets become blurred. For example, it is not always easy to tell when ramets come from the same plant, either because connections may not be clearly visible (for example when subterranean rhizomes connect apparently separate shoots) or because connections may break, isolating clonal fragments of the same individual. It is not at all clear that genets are the best population units to measure in these species. Single ramets are potentially capable of independent lives, having leaves, a stem node from which roots can be produced, and a growing point as their constituent parts. Once established and independent they behave like genetically distinct individuals, competing with neighbouring genets and other clonal fragments. For this reason, and for ease of recording, a decision is often made when censusing clonal plant species to record ramets rather than genets. Ramets may be tillers in grass species and shoots or rosettes in dicotyledonous herbs.

The use of different types of units as a basis for population studies has been discussed by Harper and White 1974; Bazzaz and Harper 1977; Hunt 1978; Harper 1977, 1978, 1981a,b 1985 and White 1979, 1980).

4.4 CASE STUDIES INVOLVING MONITORING OF RARE PLANT SPECIES

In this section, some published studies of plant populations are used to illustrate the value of this type of research in the conservation of rare and endangered plant species. All of the studies referred to have involved censuses, and in addition to presence and absence data, information about the performance of individual plants was recorded. In the accounts which follows, emphasis is placed on the results obtained rather than on the methodology.

4.4.1 Early spider orchid, *Ophrys sphegodes*

Ophrys sphegodes, the early spider orchid, is a species which has declined sharply in its range in the British Isles in the last 50 years. In a long-term study of this species, Hutchings (1987b) showed that sexual maturity was achieved in the first year above ground by about 70% of plants, that nearly all recruitment was via seed and that vegetative proliferation was rare. In addition, the half-life of orchids in the study population was only about two years, measured from the date of first appearance above ground. Half-life is a statistic used by demographers to measure the time it takes for half of a cohort of plants (all the plants starting life within a short time interval), to die. It is identical in meaning and calculation to the half-life of a radioisotope. Half-lives have been calculated for many herbaceous species, and they range from weeks or months for annuals to decades or even centuries for perennials. Half-lives in the Orchidaceae might be expected to be long, diminishing the frequency of running the gauntlet of the substantial risks associated with establishment from seed. In *O. sphegodes*, the remarkably short half-life results in rapid flux of plants through the population (Hutchings 1987a) and, if recruitment fails, or stays at a low level, a considerable risk of quick extinction.

During the first half of the study on *O. sphegodes* (1975–9), cattle grazed the site on which the orchids grew. In 1980, the cattle were replaced by sheep, and from 1981 onwards sheep grazing was only permitted when the orchid was not flowering or maturing seeds. Grazing is beneficial to *O. sphegodes* because it maintains a short turf, thus reducing interspecific competition.

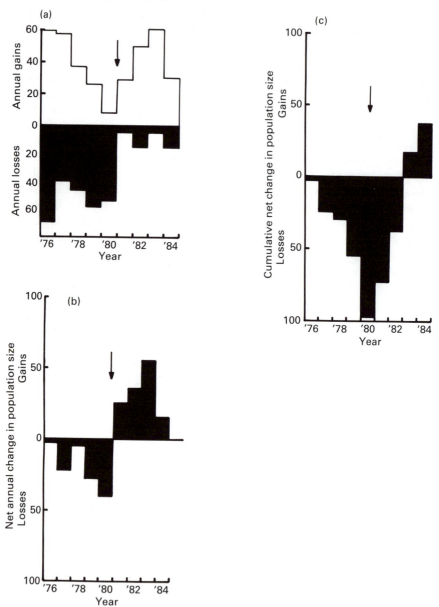

Figure 4.1 Flux analysis in a population of *Ophrys sphegodes*, 1975–84. (a) Annual recruitment (□) and annual mortality (■). (b) Annual net yearly change in the size of the population. (c) Cumulative net yearly change in the size of the population. Arrows mark the change in management from cattle grazing (1975–9) to sheep grazing (1981–4). From Hutchings (1989b).

Sheep grazing is more favourable than cattle grazing, because sheep crop the vegetation closer to the ground, and cause less mechanical damage to the soil and the plants within it.

The change in grazing regime coincided with, and is probably responsible for, considerable changes in the population biology of *O. sphegodes*. The effects can be seen in Figure 4.1. During the years of cattle grazing, annual recruitment rapidly fell and annual mortality was high (Figure 4.1a). The yearly change in the size of the population was always negative (Figure 4.1b), and the population had an accumulated deficit, from 1975 to 1980, of nearly 100 more deaths than births (Figure 4.1c). From 1980 onwards the situation altered dramatically. Annual recruitment rose rapidly, annual mortality almost ceased (Figure 4.1a), and the net annual change in population size became positive (Figure 4.1b), so that by 1984 the population had achieved substantially more recruitments than deaths over the whole ten year period of the study. This immediate change from an excess of deaths during the period of cattle grazing to an excess of births when sheep grazing was introduced, could not be deduced from the counts of population size (Figure 4.2). It could only be established by analysing the population flux from the census data. It is also the case that without the census data the presence of dormant plants in the population could not be established. Recognition of this component of the population resulted in a doubling of its estimated size (Hutchings 1987a). Thus, the analysis of census data allowed rapid verification that the new management regime imposed on this species was beneficial.

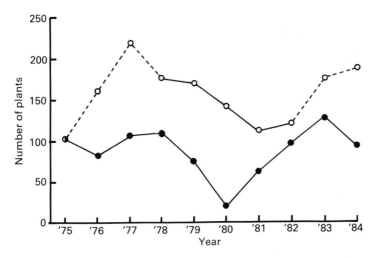

Figure 4.2 Number of emergent plants (●) and estimated total population size including dormant plants (○) in a population of *Ophrys sphegodes* from 1975–84. From Hutchings (1990).

4.4.2 Endemic sand dune species

The lengthy study of *O. sphegodes* provided information which, as time passed, was used to modify management recommendations. Conservation of *O. sphegodes* is an objective which we should be able to achieve. Studies of such length are not always necessary to establish the status of populations or to enable a management plan to be developed. A detailed two-year study of three rare endemic sand dune perennial species by Pavlik and Barbour (1988) provided enough information to demonstrate that the populations of these species were either stable in size or growing, and Pavlik and Barbour concluded that apart from ensuring adequate protection from human disturbance, the three species had demographic properties which should ensure their persistence. All that was required in the way of future research was occasional monitoring (by using this term, as indicated earlier, they imply *censusing*) to verify that their conclusions still applied. Resources could therefore be concentrated on studying more critically threatened species. Rarity does not therefore necessarily mean that management must be altered to safeguard the future of plant species.

4.4.3 Plantago cordata

Another species which has suffered a marked contraction in range is *Plantago cordata*. It used to be common in the eastern United States, but now occurs in few, widely scattered localities. It is listed as an endangered species by the US Department of the Interior. Meagher, Antonovics and Primack (1978) carried out an exceptionally broad and valuable study on this species, involving, amongst many other lines of investigation, the collection of census data over four flowering seasons. From these data they conducted an analysis of the behaviour of individual plants in consecutive years of the study. Both leaf number and number of inflorescences produced by plants showed a general decline from one year to the next (Figure 4.3). This suggests that the species could be in serious decline, since plants which survived to become a year older might be expected to improve, rather than deteriorate in performance. In addition, even though the survivors to the second year were growing at a lower density because of mortality in the populations, the proportion of flowering plants fell from one year to the next. (In *P. cordata*, as in *Isotria medeoloides* [Mehrhoff 1989], the flowering plants were the largest in the population. Mehrhoff has also shown that the largest, flowering plants have the highest survivorship. If, as we might expect, this also applies to *P. cordata*, the fall in the proportion of plants flowering raises still greater worries about the continued survival of this species.) Seedling establishment, both where *P. cordata* occurred naturally and in favourable sites into which adult plants were successfully transplanted, is extremely rare (Meagher *et al.*

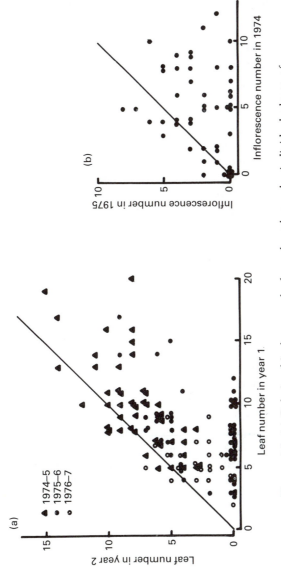

Figure 4.3 (a) Relationship between leaf numbers borne by individual plants of *Plantago cordata* in two consecutive years of a demographic study. (b) Relationship between inflorescence number borne by individual plants of *Plantago cordata* in 1974 and 1975. From Meagher, Antonovics and Primack (1978).

1978), so that this recruitment from seed cannot be relied on to maintain populations.

Plantago cordata populations are found on gravel bars in woodland river channels. Whereas the species is well adapted to stable stream environments of climax hardwood forests, many of these habitats have been destroyed or significantly altered by man's activities, including drainage, ditching, damming and re-routing of streams, and the clearance of surrounding woodlands. These activities are the cause of its decline. Although *P. cordata* can grow in full sunlight, so that woodland clearance is not directly detrimental, the habitats which it can occupy are more prone to flooding, scouring and erosion because of these disturbances, resulting in dislodging of adult plants and failure of seedlings to establish and survive to adult status. *P. cordata* has extensive fibrous root systems (which are adequate for anchorage in natural habitats but not in disturbed habitats) and large leaves to maximise light interception in its natural shady habitats. These features are combined with the lowest reproductive effort (calculated both as the number and weight of seeds produced per unit area of leaf) measured for any species within the large genus *Plantago*. The evolution in the past of a pattern of high allocation of resources to leaves and roots, limits the extent to which a greater reproductive effort could evolve in response to its present needs. Therefore, there seems little prospect of the species increasing its seed output or of improving its frequency of establishment from seed in the unstable environments created by human influence. The shortage of undisturbed habitats clearly reduces the prospects for conserving natural populations of the species.

Fortunately for the survival of *P. cordata*, it can easily be conserved by growing it in greenhouses and botanical gardens. It can also be successfully established in many places in the wild, at least in the short term. Pleasingly, Meagher *et al.* (1978) make a persuasive case for conserving this species, which may come as something of a surprise to readers who regard the genus *Plantago* as neither attractive, interesting nor valuable. The attitude of Meagher *et al.* is laudable and it should be maintained towards all of our endangered species, regardless of the affection in which they are held. Although all have taken millions of years to evolve, we have the power to extinguish them in years or decades. It is our duty, however, to work for the conservation of all of them.

4.5 TOWARDS MORE MONITORING STUDIES ON PLANT SPECIES

'Plants stand still and wait to be counted' (Harper 1977, p. 515). It would be unwise to take the implications of this oft-quoted comment strictly at face value; Harper, the catalyst who sparked off so much of the present-day interest in demographic studies on plants, was undoubtedly urging the study of plant population biology by census. Notwithstanding the many difficulties in repeatedly counting and measuring the same individual plants in the field, there is rarely a good excuse for undertaking counts rather than censuses, especially when the populations present management problems. Inventories, surveys and counts are only preliminary methods of data collection; they have little to offer when we want to know how to manage species. The need to commission more census studies of threatened species is recognised by many nature conservationists, but inventory and survey still attract the most funding, and the interested amateurs who offer their services to conservation organisations are all too often encouraged to count plants, when better use could be made of their enthusiasm. In short, there is little that can be done with counts of plants; this is emphasised by the fact that long runs of count data are rarely published in any form.

In general, when habitats and communities are conserved, there is less threat of extinction to their component species. We should therefore accept that habitat and community conservation usually take priority over species conservation. While many would also accept that rare species in communities are usually more in need of sympathetic management than common species, we can only produce rational plans for managing or conserving *any* species after we understand its ecology. Several valuable publications, all on rare species, contain this message in one form or another. Griggs and Jain (1983) state that 'only after the population structure and life history of each species are known adequately can species-specific decisions be made toward optimal conservation strategies'. Bratton (1985) comments that 'our ability to protect or manage many herbaceous species is limited by a lack of demographic information, and a poor knowledge of how different types of herbs respond to disturbances, including grazing and other sorts of predation. More studies quantifying factors causing population declines are badly needed.' Finally, Pavlik and Barbour (1988) express the opinion that 'preservation management requires a knowledge of the population's status (whether it is declining, stable or growing) under the most "natural" of conditions. This determination can be made by selecting a few, relevant population attributes and conducting a demographic monitoring programme'. The same point emerges most vividly from Mehrhoff's (1989) study of tiny, declining populations of the endangered orchid *Isotria medeoloides*.

All of the papers quoted above are by North American writers. At present, conservationists in the UK appear to lag far behind, either in realising the importance of what these authors are saying, or failing to act upon it. In 1974 the Nature Conservancy Council in the UK financed a project to *survey* rare plants in eastern England, a project regarded as 'a major shift in emphasis, from recording only the presence or absence of species in 10 km grid squares ... to a recognition of the need to record the exact location of rare species and evaluate conservation measures for their survival' (Crompton 1981). While this was certainly an important and welcome step in the right direction, the method of data collection was still far removed from the ideal of censusing, so that the shift in emphasis still moved only half-way to the goal – to provide information to secure the future of populations of rare and threatened plants. While this may be best achieved through enlightened habitat and community management, we can only know whether our management is good or bad for a species, if we have analysed its demographic properties, and the ways in which these are altered by changes in management. In view of the regrettable scarcity of such analyses it seems appropriate to end with a plea for more funding for this type of research.

REFERENCES

Ayensu, E.S. (1981) Assessment of threatened plant species in the United States, in *The Biological Aspects of Rare Plant Conservation* (ed. H. Synge), Wiley, Chichester, pp. 19–58.

Bazzaz, F.A. and Harper, J.L. (1977) Demographic analysis of the growth of *Linum usitatissimum*, *New Phytologist*, 78, 193–208.

Bradshaw, M.E. (1981) Monitoring grassland plants in Upper Teesdale, England, in *The Biological Aspects of Rare Plant Conservation* (ed. H. Synge), Wiley, Chichester, pp. 241–51.

Bratton, S.P. (1985) Effects of disturbance by visitors on two woodland orchid species in Great Smoky Mountains National Park, USA, *Biological Conservation*, 31, 211–27.

Bullard, E.R., Shearer, H.D.H., Day, J.D. and Crawford, R.M.M. (1987) Survival and flowering of *Primula scotica* Hook, *Journal of Ecology*, 75, 589–602.

Crompton, G. (1981) Surveying rare plants in Eastern England, in *The Biological Aspects of Rare Plant Conservation* (ed. H. Synge), Wiley, Chichester.

Davy, A.J. and Jefferies, R.L. (1981) Approaches to the monitoring of rare plant populations, in *The Biological Aspects of Rare Plant Conservation* (ed. H. Synge), Wiley, Chichester, pp. 219–32.

De Steven, D. (1989) Genet and ramet demography of *Oenocarpus mapora* ssp. *mapora*, a clonal palm of Panamanian tropical moist forest, *Journal of Ecology*, 77, 579–96.

Greig-Smith, P. (1983) *Quantitative Plant Ecology* (3rd ed.), Blackwell, Oxford.

Griggs, F.T. and Jain, S.K. (1983) Conservation of vernal pool plants in California, II. Population biology of a rare and unique grass genus *Orcuttia*, *Biological Conservation*, 27, 171–93.

Harper, J.L. (1977) *Population Biology of Plants*, Academic Press, London.

Harper, J.L. (1978) The demography of plants with clonal growth, in *Structure and Functioning of Plant Populations* (eds A.H.J. Freysen and J.W. Woldendorp), North-Holland Publishing Company, Amsterdam, pp. 27–48.

Harper, J.L. (1981a) The meanings of rarity, in *The Biological Aspects of Rare Plant Conservation* (ed. H. Synge), Wiley, Chichester, pp. 189–203.

Harper, J.L. (1981b) The concept of population in modular organisms, in *Theoretical Ecology: Principles and Applications* (2nd ed.) (ed. R.M. May), Blackwell, Oxford, pp. 53–77.

Harper, J.L. (1985) Modules, branches, and the capture of resources, in *Population Biology and Evolution of Clonal Organisms* (eds J.B.C. Jackson, L.W. Buss and R.E. Cook), Yale University Press, New Haven, pp. 1–33.

Harper, J.L. and White, J. (1974) The demography of plants, in *Annual Review of Ecology and Systematics*, 5, 419–63.

Hunt, R. (1978) Demography versus plant growth analysis, *New Phytologist*, 80, 269–72.

Hutchings, M.J. (1986) Plant population biology, in *Methods in Plant Ecology* (2nd ed.) (eds P.D. Moore and S.B. Chapman), Blackwell, Oxford, pp. 377–435.

Hutchings, M.J. (1987a) The population biology of the early spider orchid, *Ophrys sphegodes* Mill. I. A demographic study from 1975–1984, *Journal of Ecology*, 75, 711–27.

Hutchings, M.J. (1987b) The population biology of the early spider orchid, *Ophrys sphegodes* Mill. II. Temporal patterns in behaviour, *Journal of Ecology*, 75, 729–42.

Hutchings, M.J. (1989a) Population biology and conservation of *Ophrys sphegodes*, in *Modern Methods in Orchid Conservation – The Role of Physiology, Ecology and Management*, (ed. H.W. Pritchard), Cambridge University Press, Cambridge, pp. 103–17.

Hutchings, M.J. (1989b) Conservation and the British orchid flora, *Plants Today*, 2, 50–8.

Hutchings, M.J. (1990) The role of demographic techniques in conservation – the case of *Ophrys sphegodes* in chalk grassland, in *Calcareous Grasslands – Ecology, and Conservation*, (eds S. Hillier, D. Wells and D.W.H. Walton), *in press*.

Jeffers, J.N.R. (1978) *Design of Experiments*. Statistical Checklist 1. Institute of Ecology, Cambridge.

Jeffers, J.N.R. (1979) *Sampling*. Statistical Checklist 2. Institute of Ecology, Cambridge.

Kershaw, K.A. and Looney, J.H. (1985) *Quantitative and Dynamic Plant Ecology* (3rd ed.), Edward Arnold, London.

Leverich, W.J. and Levin, D.A. (1979) Age-specific survivorship and reproduction in *Phlox drummondii*, *American Naturalist*, 113, 881–903.

Meagher, T.R., Antonovics, J. and Primack, R. (1978) Experimental ecological genetics in *Plantago*. III. Genetic variation and demography in relation to survival of *Plantago cordata*, a rare species, *Biological Conservation*, 14, 243–57.

Mehrhoff, L.A. (1989) The dynamics of declining populations of an endangered orchid, *Isotria medeoloides*, *Ecology*, 70, 783–6.

Moore, P.D. and Chapman, S.B. (1986) *Methods in Plant Ecology* (2nd ed.), Blackwell, Oxford.

Palmer, M.E. (1987) A critical look at rare plant monitoring in the United States, *Biological Conservation*, 39, 113–27.

Pavlik, B.M. and Barbour, M.G. (1988) Demographic monitoring of endemic sand dune plants, Eureka Valley, California, *Biological Conservation*, 46, 217–42.

Perring, F.H. and Farrell, L. (1983) *British Red Data Book, 1. Vascular Plants* (2nd ed.) Society for the Promotion of Nature Conservation, Lincoln.

Prince, S.D. (1986) Data analysis, in *Methods in Plant Ecology* (2nd ed.) (eds P.D. Moore and S.B. Chapman), Blackwell, Oxford, pp. 345–75.

Rabinowitz, D. (1981) Seven forms of rarity, in *The Biological Aspects of Rare Plant Conservation* (ed. H. Synge), Wiley, Chichester, pp. 205–17.

Roberts, H.A. (1981) Seed banks in soil, *Advances in Applied Biology*, 6, 1–55.

Tamm, C.O. (1972) Survival and flowering of some perennial herbs. II. The behaviour of some orchids on permanent plots, *Oikos*, 23, 23–8.

Thompson, K. (1986) Small-scale heterogeneity in the seed bank of an acidic grassland, *Journal of Ecology*, 74, 733–8.

Watkinson, A.R. (1978a) The demography of a sand dune annual: *Vulpia fasciculata* II. The dynamics of seed populations, *Journal of Ecology*, 66, 35–44.

Watkinson, A.R. (1978b) The demography of a sand dune annual: *Vulpia fasciculata* III. The dispersal of seeds, *Journal of Ecology*, 66, 483–98.

Wells, T.C.E. (1981) Population ecology of terrestrial orchids, in *The Biological Aspects of Rare Plant Conservation* (ed. H. Synge), Wiley, Chichester, pp. 281–95.

White, J. (1979) The plant as a metapopulation, *Annual Review of Ecology and Systematics*, 10, 109–45.

White, J. (1980) Demographic factors in populations of plants, in *Demography and Evolution in Plant Populations* (ed. O.T. Solbrig), Blackwell, Oxford, pp. 21–48.

White, J. (1985a) The census of plants in vegetation, in *The Population Structure of Vegetation. Handbook of Vegetation Science, Part III* (ed. J. White), Dr. W. Junk, Dordrecht, pp. 33–88.

White, J. (ed.) (1985b) *The Population Structure of Vegetation. Handbook of Vegetation Science, Part III*, Dr. W. Junk, Dordrecht.

White, P.S. and Bratton, S.P. (1981) Monitoring vegetation and rare plant populations in US national parks and preserves, in *The Biological Aspects of Rare Plant Conservation* (ed. H. Synge), Wiley, Chichester, pp. 265–78.

Williams, O.B. (1981) Monitoring changes in populations of desert plants, in *The Biological Aspects of Rare Plant Conservation* (ed. H. Synge), Wiley, Chichester, pp. 233–40.

— 5

Vegetation monitoring

BARRIE GOLDSMITH

5.1 WHAT IS VEGETATION?

Ecologists frequently focus on vegetation for their monitoring studies. But why? The answer is probably due to its relative permanence, at least compared with many groups of animals and relative ease of recording. Also because its species composition reflects the nature of underlying soils, local climate, current and past management and is the medium in which animals live and feed. This wide range of controlling factors makes it difficult to interpret the requirements of the species that we record although the more we learn of the ecology of individual species the easier it becomes to interpret the vegetation of a particular area.

Plant ecologists can aim to study the requirements of individual species (autecology) or of communities (synecology). Individual species have relatively precise requirements, but communities, being an imprecise concept, have a range of controlling factors which renders them unsuitable for most monitoring activities.

This chapter deals with natural or semi-natural terrestrial vegetation which consists of self-sown plants, sometimes consisting of patches or discrete classes but more often a continuous gradient of different species. Subdivision into classes has frequently been the goal of ecologists and can have value in some studies. Continental ecologists have favoured using subjective classificatory techniques to describe vegetation but they are not very precise and are therefore not ideal for monitoring purposes.

Vegetation consists of large numbers of individual plants which can be grouped together into species but the species usually cannot be grouped into classes with clear boundaries. If boundaries are apparent they are usually due to environmental or management factors and it may be easier to record these rather than the vegetation responding to them. This is why classificatory techniques are not suitable and this criticism is as valid for the more mathematical, objective methods of classification as for the subjective ones.

So classificatory or phytosociological techniques are not ideal for monitoring and we must look for something more precise such as recording the abundance, or changing abundance of individual species, or a selection of species. We must be able to position samples in the field very precisely, be able to relocate them (which is easier said than done), and determine change due to the variable that interests us over short periods of time, simply because funding constraints do not usually permit long-term monitoring. However, vegetation and its component species are changing due to a wide variety of factors such as:

- seasonal effects
- long-term directional changes or succession
- management, e.g. manipulating water-levels with sluices, culling rabbits
- natural cycles, such as those of grazing herbivores and predators
- climatic change or its consequences.

A dilemma occurs when the person carrying out the monitoring is focusing on one of the above but that type of change is, or could be, affected by one or more of the other types of change. The research worker may be unaware of which other factors are operating or that others may become operative during the monitoring exercise. For example we may wish to study the effects of long-term global changes, but the effects of local management may be much greater and obscure the underlying and longer-term trends.

Synecology is attractive because it is holistic but it suffers the disadvantage that it is imprecise. So should the reader abandon monitoring vegetation now and focus on monitoring single species in which case the chapter on monitoring rare plants (Chapter 4) would be the appropriate one to read? The answer will depend on the circumstances and will often be 'No', but he should avoid phytosociological techniques and anything involving vegetation boundaries unless there are special reasons for including them. It will often be better to concentrate on the precise recording of samples with precise measures of abundance of all or selected species.

5.2 SAMPLES

It is clearly impractical to record everything, so the plant ecologist usually locates a limited number of samples. How these are located will be discussed later. The word sample is perfectly acceptable but ecologists have used a variety of other words, often with imprecise and overlapping meanings. These include the following:

- **stand:** which originally came from forestry and has a precise meaning, usually a square area of uniform vegetation for recording; widely used in plant ecology;
- **plot:** possibly originating from agriculture, usually a square, the meaning sometimes implies harvesting;

- **relevé:** from the French and widely used in phytosociology, similar meaning to stand but usually a temporary sample and may have implications of imprecision;
- **quadrats:** these should be used to record abundance within samples and so this term should not be used in this context;
- **plotless:** a type of recording used with trees, e.g. the point-centred quarter method, which works well where appropriate.

There are no simple rules as to how many samples are necessary (see Chapter 2). The general maxim 'the more the better' holds true. For a critique of fixed quadrats, see p. 11.

5.3 LOCATION OF SAMPLES

Ideally samples should be located at random so that every part of the study area has an equal chance of being recorded. This is usually achieved by:

1. Defining the study area and if necessary marking it out.
2. Laying down two lines at right angles outside the study area. A 3.4.5 triangle made with 30 m tapes can be used to define a right angle.
3. Pairs of random numbers taken from statistical tables are then used to locate distances along the two lines (ordinate and abscissa).
4. The sample is then located where the ordinate and abscissa intercept. If samples fall outside the study area they can be rejected.
5. Recording is carried out. This usually involves determining the abundance of one or several species.
6. The position of the sample must be permanently marked if the same points are to be used again, e.g. next year.

Another procedure which has been used is the 'line transect', which is a series of sampling points along a straight line. They are particularly popular where there are obvious gradients of vegetational variation, e.g. salt-marshes, where the transect is located parallel to the gradient, in this example from the top to the bottom of the marsh. The method has the advantage that it can be relatively easily relocated and samples can be quickly located. However, the data have not been collected from randomly selected points and are therefore not amenable to statistical analysis. Also the area is not evenly sampled so the results may be heavily biased. However change is readily recorded and there are some useful transects in Britain which show nearly 40 years of vegetation change, e.g. the A. S. Thomas transects on chalk grassland which were established in the 1960s. Since the cessation of grazing some of these have developed so much scrub that they appear to be old hedges.

5.4 MEASURES OF ABUNDANCE OF SPECIES

There are many of these and they have been well reviewed in numerous textbooks, e.g. Causton (1988), Greig-Smith (1983), Kershaw and Looney (1985), Moore and Chapman (1986). Thus it is only necessary here to give some brief comments about the main methods in so far as they affect their value for monitoring.

(a) Presence or absence of species

Or species lists without any measure of abundance. This is easy to record but will not be sensitive to changes in the abundance of different species. It can be useful where vegetation is developing and species number is increasing markedly such as on shingle, wasteground, glacial moraines or volcanic larva.

(b) Presence or absence combined with frequency symbols

These are sometimes referred to as DAFOR, an acronym standing for Dominant, Abundant, Frequent, Occasional and Rare. As there are no precise definitions of these terms they are best avoided for monitoring purposes unless time is extremely limited. They are sometimes qualified with letters standing for words such as 'locally' or 'very' but these do not help very much. It is better to use one of the measures below.

(c) Absolute measures such as density, cover, biomass or basal area

Density is the number of individuals per unit area. It is widely used with animals which are usually discrete units and are therefore easily counted. Plants often spread vegetatively so making this measure difficult to use, except for certain species, such as bulbs, orchids, annuals and trees.

Cover is the proportion of ground covered by a species and should be envisaged as a vertical projection of the species on to the ground. It is usually recorded with pins of very narrow diameter located at random. Sometimes vertically-arranged pairs of cross-wires may be used as a series in a frame. Similarly the pins can be held in a frame but this will affect the randomness of the sampling. My experience is that random sampling using individual pins is the most accurate technique for monitoring but it is extremely tedious. Some workers estimate cover by eye but this can be very inaccurate, especially when different recorders are involved. Estimates of cover are often placed in ranges of value, e.g. the Domin 1 to 10 range which is used on the continent of Europe and in the British National Vegetation Classification:

Domin value	% Cover range
10	91–100
9	76–90
8	51–75
7	34–50
6	26–33
5	11–25
4	5–10
3	less than 4, frequent
2	less than 4, occasional
1	less than 4, rare

A glance at the right-hand column illustrates that this scale is based on a series of fractions. It is currently very popular in phytosociology and works well in that context. It is quick and produces satisfactory ordinations and classifications. However it is not sufficiently accurate for most monitoring exercises.

Biomass or yield A plot is usually harvested, the material sorted into species and the dry weight of each obtained. This is accurate but destructive and therefore not repeatable. Neither is it ideal for use on nature reserves or other special areas. This method is used in 'production ecology' but there are few circumstances where it is appropriate to monitoring studies.

Basal area is appropriate to trees or plants with a tussock morphology. It is similar to cover in terms of its advantages and disadvantages and is not often used in monitoring studies.

(d) Non-absolute measures
The most popular non-absolute measure is frequency. This is defined as the proportion or percentage of quadrats which contain a particular species. Non-absolute refers to the fact that values are dependent on quadrat size which should always be stated. It has the important advantage that it is very easy to use and many students first encounter it when they are at school. However, the values combine information about density and pattern which cannot subsequently be separated. For a discussion of this see the textbooks listed earlier in Section 5.4. As it is easy to use, a large number of quadrats can be recorded in a short period of time which increases the accuracy of the results. Consistency between different workers is also usually high.

(e) *Measures of performance*

These include height, growth rate, and indices of morphology. They are not very often used but should be carefully considered especially if a measure of species response is needed in relation to a small change in an environmental or management factor. Scott (1989) used 2×2 metre quadrats positioned by random numbers as a baseline for a monitoring study of Meadow Clary (*Salvia pratensis*). In each he recorded:

1. number of plants of Meadow Clary and number in flower;
2. number and type of shoot per plant;
3. number of shoots in bud or flower, and with unripe or ripe seed;
4. basal rosette diameter, height, degree of branching and number of whorls per flowering shoot;
5. amount of grazing damage per shoot;
6. number of whorls destroyed by insects and small mammals per flowering shoot;
7. average sward height ignoring emergent inflorescences;
8. dimensions of anthills, presence of small mammal runs and worm casts;
9. slope and aspect;
10. percentage cover of all vascular plant species, bryophytes and bare ground.

This shows the range of variables which can be recorded in a surveillance or monitoring exercise (for the differences between these see Chapter 1). The attributes of the flowering shoots were recorded as the mean with its standard error but these kinds of data should not be assumed to be normally or symmetrically distributed about the mean. More information about monitoring single species appears in Chapter 4.

5.5 QUADRAT SIZE

There is no magic rule for determining the optimum quadrat size, but the principles discussed in the various plant ecology texts apply here. First, the quadrat should be small enough to be searched easily, and smaller sizes take less time per quadrat thereby permitting more quadrats to be used and so increasing the accuracy of the estimates. However larger quadrats increase the area-to-edge ratio, and thereby reduce the edge effect or edge error. Quadrats should be large enough for most of them to include one or more of the largest individuals or patches. This will vary for different species but usually only one quadrat size is practical so necessitating some compromise. If frequency is being recorded most species should ideally have 20%–70% and if one or more has values of 100% the quadrat size is probably too big. With density the number of plants per quadrat should be easily countable and the use of excessive numbers of empty quadrats is inefficient and suggests that the quadrat size should be increased. It is worth bearing in mind that the next time the recording is carried out the species may be either less or more abundant but that the quadrat size cannot be changed during a monitoring project (p. 220).

5.6 SAMPLING PATTERNS

There are three main types of arrangement of vegetation samples: random, systematic and stratified.

(a) Random

With random sampling every point in the study area has an equal chance of being sampled. The data are not biased, and are suitable for statistical analysis. The location of random points has been described above. It takes a little time but for monitoring purposes the time is well spent. It is only in certain circumstances, e.g. dense forest or sticky mudflats, that some other sampling system should be contemplated. Some authorities recommend a 'random' walk but this is not random and is not recommended. If time is limited it is acceptable to pace out the location of random points derived from random coordinates.

(b) Systematic

This type of sampling usually involves a grid and the points may therefore be easier to locate and relocate. This is a considerable advantage for a monitoring exercise. However no measure of the accuracy or reliability of the measurements can be determined and no statistical analysis of the results can be carried out (see later discussion on comparing two sets of data). The worst scenario would occur where the pattern of the sampling grid coincided with the pattern in the site or the vegetation. Examples of this include sampling broad ridge and furrow or tree crops such as orchards.

(c) Stratified

This type of sampling occurs when the study area is divided into zones using topography and then each is sampled at random, usually, but not necessarily, with a number of samples proportional to the area of each zone. Subdivision based upon vegetation types is often used but not recommended because of the risk of circularity of argument and the fact that the zones are often imprecise and may change during the period of the monitoring exercise. Stratified random sampling incorporates the advantage of systematic sampling which ensures a more even spread of samples across the area and may thereby increase the accuracy of the results. Fewer samples can be located in less important or less interesting areas if required thereby saving time. The accuracy of estimates from these areas will however be lower as a result.

5.7 MAPPING SCHEMES

Because of the absence of natural vegetation boundaries, monitoring schemes should aim to avoid mapping exercises. There are a series of atlases of flora at national and county scales which may appear ideal monitoring exercises. For more information about these see Chapter 8, by Paul Harding. However they are not ideal monitoring schemes because of the difficulty of recording when species disappear and the lack of sensitivity when rerecording the information. The data represent the accumulation of many years data and not a single snapshot in time which could be taken again, say, a decade later. We can however learn a considerable amount about how to (and how not to) record the location of samples and processing data from these schemes. In this regard some, such as the *Flora and Vegetation of County Durham* (Graham 1988) are very useful.

The first comprehensive mapped Flora was the *Atlas of the British Flora*, which involved data collected by volunteers between 1954 and 1960 (Perring and Walters 1962). It involved 1½ million records and shows the distribution of 1700 species. The information is presented at the scale of 10×10 km squares or 100 sq km which is not very precise but shows general geographical, climatic and geological trends and relationships. However records are only shown for two main time periods, pre-1930 and post-1930, which are not comparable with each other in terms of the amount of data available and are too imprecise for any but the most general monitoring purposes (see p. 150).

The *Flora of Hertfordshire* (Dony 1967) covers a single English county and is based on smaller squares, each 2×2 km or 4 sq km known as 'tetrads'. These are four of the small squares bounded by even-numbered kilometre lines of the national grid as marked on Ordnance Survey maps. Thus there are 25 per 10×10 km square. They are numbered from A to Z (omitting O) from the bottom left-hand corner. This Flora involved 117 703 records for 335 complete squares and 172 partial ones. There were on average 261 species per tetrad with 70 species occuring in all tetrads. This is a more detailed scheme at a finer scale but there is no information mapped relating to time. Therefore it is not a good baseline for monitoring. This Flora also has 109 habitat lists each with a grid reference but it is not sufficient information to enable precise rerecording and cannot therefore be seen as an ideal monitoring exercise.

There have been many other County Floras, some with more detailed information, for example *The Flora of Essex* (Jermyn 1974) deals with 1733 species and states 'that 10 km is too large a unit to give a reasonable assessment of frequency' and uses 3988 individual 1×1 km squares. Each is

referenced by using the standard letter code for the 10 × 10 km squares, e.g.
TL, and then two figures are used to give an 'easting' read from the grid line
coordinates to the south and two figures are read to give a 'northing' read
from grid line coordinates to the east. The level of detail here, together with
material in the text, is impressive but whether this forms the perfect baseline
for monitoring is debatable. Even within a grid square a species may be
occupying several different sites and a considerable amount of change could
occur before the distributions, as mapped, would alter. Even so it is debatable
when County Floras will be remapped rather than updated.

5.8 COMPARING TWO SETS OF DATA

Having collected data for a site at one time and then again after a few years
the ecologist carrying out the monitoring probably wishes to establish
whether or not the data are genuinely different. Let us consider an example.
If the mean density of a species on two occasions increased from ten to 19 one
might be tempted to suggest that a real increase had occurred. If, however,
the standard errors of the means were three and four respectively, and the
assumptions of parametric statistics were met, we could then calculate the
appropriate statistics. We could determine the standard error of the
difference of the means and the appropriate t value of 1.8 (see below). The
corresponding value of t at the 5% level is approximately two and therefore
the probability of obtaining such a large difference by chance would be more
than 5%. Our considered judgement, aided by these simple statistics must be
that no real change had occurred.

	A	B
mean density	10	19
SEM	3	4

$$\text{SE diff} = \text{sq. rt. } (3^2 + 4^2) = 5$$
$$t = \text{obs. diff.}/\text{SED} = 9/5 = 1.8$$
$$t_{5\%} = \text{approx. } 2$$

Therefore the chance of getting such a large difference by chance would be
more than 5%. Comparing two samples in this way is acceptable but repeated
sampling at the same position leads to the problem of autocorrelation and
should be avoided (see Chapter 2).

5.9   CONCLUSION

When monitoring vegetation it is preferable to focus on samples which have
been selected in a random or stratified random manner and to record a precise
measure of abundance such as cover. Attempts to document boundaries are
not recommended, neither are techniques based on classification or
phytosociology as they are too imprecise.

It may be difficult to monitor a long-term underlying trend due to say climatic change if shorter-term management-induced changes such as changes in grazing regime have greater impact. This can cause considerable problems even where reserves have been selected for major monitoring exercises, for example, to determine the effects of important phenomena such as sea-level rise.

REFERENCES

Causton, D. (1988) *Introduction to Vegetation Analysis*, Unwin, London.

Dony, J.G. (1967) *Flora of Hertfordshire*, Hitchin Museum, Hitchin.

Graham, G.G. (1988) *The Flora and Vegetation of County Durham*, Durham Flora Committee and Durham County Conservation Committee, Durham.

Greig-Smith, P. (1983) *Quantitative Plant Ecology*, Blackwell, Oxford.

Jermyn, S.T. (1974) *Flora of Essex*, Essex Naturalists Trust, Colchester.

Kershaw, K.A. and Looney, J.H.H. (1985) *Quantitative Plant Ecology*, Arnold, London.

Moore, P.D. and Chapman, S.B. (1986) *Methods in Plant Ecology*, Blackwell, Oxford.

Perring, F.H. and Walters, S.M. (1962) *Atlas of the British Flora*, Botanical Society of the British Isles, Nelson, London.

Scott, A. (1989) The Ecology and Conservation of Meadow Clary *Salvia pratensis*, M.Sc. Dissertation, University College, London.

6

Monitoring butterfly numbers

ERNEST POLLARD

6.1 INTRODUCTION

Butterflies are attractive, colourful and they fly by day. These characteristics not only make them amongst the most widely known groups of insects, and certainly the most popular, but also very suitable for monitoring. There is widespread interest in the study and conservation of butterflies and so an abundance of potential recorders who can readily identify them.

Four butterfly species have become extinct in the UK in the last 150 years. There are now 58 species of resident and common migrant butterflies in the UK and of these 18 have declined sharply in their ranges during the present century (Heath *et al.* 1984). The main reason for these declines is thought to be the change from traditional methods of agriculture and forestry to monocultures in the form of weed-free grasslands, forests and crops. Several butterfly species are now largely restricted to nature reserves and other protected areas where older forms of land-use are maintained.

A scheme for monitoring the abundance of butterflies began in 1976. The aims of this scheme are:

1. To provide information at regional and national levels on changes in butterfly numbers and to detect trends which may affect the status of butterflies.
2. To monitor changes in numbers at individual sites and, partly by comparisons with results elsewhere, assess the impact of local factors such as habitat change.

Monitoring itself provides no information on the causes of the changes recorded, but nevertheless it may be possible to interpret the patterns of change and so contribute to the understanding of the population ecology of butterflies. For example, studies have been made of the association between weather and butterfly numbers, as shown by data from the monitoring scheme (Pollard 1988).

In addition to the main aims, the scheme has provided a range of information on phenology and migration and also on the local distribution of butterflies along the recording routes, in a range of habitat types.

This chapter is in three parts: in the first a general account of the scheme is given, with examples of the various types of data obtained. The second part is a case study, using the results from a particular site, Castle Hill National Nature Reserve in Sussex, to illustrate the ways in which monitoring may be used to assess the impact of local habitat change. Finally, the extent to which the stated aims have been achieved is discussed.

6.2 THE BUTTERFLY MONITORING SCHEME

6.2.1 Sites

The number of sites in the scheme increased from 35 in 1976, the first year, to 77 in 1979 and has subsequently remained at about 80. Initially, the selection of sites depended largely on the availability of people willing to record; subsequently there has been some attempt to achieve a spread of habitat types and of geographical coverage. However, the sites are not at all representative of the British countryside. Nature reserves of various sorts make up about three-quarters of the sites and, geographically, there is a strong bias towards the south of England (Figure 6.1). Continuity of recording at individual sites has been good. By 1988, ten or more years' data had been recorded at 67 sites.

6.2.2 Methods

The methods used in the scheme were first tested in Monks Wood National Nature Reserve in 1973. Before the start of the national scheme in 1976, there were a further two years of pilot trials, involving six recorders in eastern England. An account of the first ten years of the national scheme is given by Pollard *et al.* (1986), including references to earlier publications on the methods and on the development of the scheme.

At each site in the scheme counts are made along a fixed route in each of 26 recording weeks from the beginning of April until the end of September each year. Selection of the route is by the recorder or, in the case of nature reserves, by the reserve manager. The route is chosen partly to be representative of the habitat types at the site, but is often restricted to areas where butterflies are known to fly. Thus, in woods, the routes are predominantly or exclusively along the rides (tracks) where butterflies occur, rather than through dense forest compartments. The forest compartments are the predominant habitat type, but usually contain virtually no butterflies. In grassland sites, this problem may not exist. The routes are divided into sections, the divisions usually coinciding with changes in vegetation type, and separate counts are made in each section.

Counts are made only when weather conditions meet specified criteria. The recorder walks at a uniform pace and records all butterflies seen within

Butterfly Monitoring
Scheme sites 1989

Figure 6.1 Location of sites in the Butterfly Monitoring Scheme in 1989.

prescribed limits. Along tracks the boundaries are often clear. The width is variable, but most are of the order of 5 metres. If no established paths are available, it may be necessary to gauge distances by eye. Various 'rules' ensure that recording is standardised as much as possible. A standard recording form (Figure 6.2) is used, so that all relevant information is given and can be quickly entered on to a computer.

Most British butterflies are identifiable in flight. However, it is recommended that recorders carry a net and catch some individuals of difficult species groups, to check identifications. One pair of species, the small and Essex skippers, *Thymelicus sylvestris* and *T. lineola*, are impossible to distinguish in flight and the data are combined.

The weather criteria were chosen as a compromise between the ideal and the practicable. If the requirements are too demanding, in terms of sunshine and temperature, then counts may not be possible in many weeks; if they are too low, counts may be attempted but few butteflies seen. In practice, for

BUTTERFLY CENSUS

YEAR	1984		DATE	10·8		RECORDER	E. Pollard
1-2			3 5			6-8	

		SITE NAME	Monks Wood		
9-11		12-17			18-19

START TIME	11·40	END TEMP °C	18·5	% SUN	70	END WIND SPEED	3
20-23		24-26		27-28		29	

SECTION		1	2	3	4	5	6	7	8	9	10	11	12	13	14	15	TOTAL
BRIMSTONE	54	4				1						3		1			9
COMMON BLUE	106																
GREEN-VEINED WHITE	99	7	2		2	1	2			1	6	2	1	2			26
HEDGE BROWN	76	2	18	4	1	14	4	11	14	16	13	12	2	44	1		156
LARGE SKIPPER	88																
LARGE WHITE	98									1		2					3
MEADOW BROWN	75		2			2	7		3	9	3	6	1	10			43
ORANGE TIP	4																
PEACOCK	84		5			1	1				2	4	1	15			29
RED ADMIRAL	122																
RINGLET	8	1	3	1		4	1	4	2	2	3	1		4			26
SMALL COPPER	68																
SMALL HEATH	29					1											1
SMALL SKIPPER	120													1			1
SMALL TORTOISESHELL	2																
SMALL WHITE	100		1				4		1			2		1			9
WALL	94		2														2
Speck. Wood		1				2			1		1						5

SECTION	1	2	3	4	5	6	7	8	9	10	11	12	13	14	15
SUNSHINE	S	S	S	S	S	S	-	-	S	S	-	-	-	S	

NOTES:

PLEASE TOTAL EACH SQUARE

most species, the chosen criteria seem satisfactory. Individual counts are usually strongly correlated with independent estimates of abundance but scarcely at all with the weather conditions during the counts (Pollard *et al.* 1986).

The weekly counts are used to calculate an index of abundance. If, as is usual, there is one count in each recording week, the index is the sum of the weekly counts. If there are two or more counts per week, the sum of the mean weekly counts is used. An index of abundance is calculated for each flight period, provided these are discrete. For species in which flight periods overlap substantially, as for example in the case of the speckled wood *Pararge aegeria*, a single annual index is given. When counts have not been made in one week, an estimate is made of the missing value; the estimate is simply the mean of the counts before and after the missing value. If more than one count is missing, a decision has to be made as to whether the calculation of an index is reasonable. For example, missing data very late in the flight period would be expected to have little impact on the size of the index and in such a case an index might be calculated even though counts for two weeks were missing.

The index values from individual sites are collated to produce regional and national index values for 29 of the commoner butterflies. The method of ratio estimates (Cochran 1963) is used. The ratio estimate is based on sites which have been recorded in consecutive years, and is simply the sum of the index values in year two divided by the sum of index values in year one. The collated index values for all species start with an arbitrary value of 100 in 1976 and subsequent values are obtained by multiplying by successive ratio estimates. The method is not without problems, in particular the collated index values may be unduly influenced by one or a few very large populations. There are alternative methods of collating individual site index values and it is possible that, in the long term, another method may prove more satisfactory for some or all species (see Chapter 12).

6.2.3 Information from butterfly monitoring

(a) Phenology

The dates of seasonal events, such as the first frog-spawn or the first cuckoo of the year, have fascinated natural historians for many years. The monitoring scheme provides an abundance of such data, not only of first sightings, which may not in fact be of great significance, but of seasonal variation in the whole flight period of butterflies.

As might be expected, most butterflies emerge earlier in warm seasons than in cold. Brakefield (1987) used data from the monitoring scheme to study

---

Figure 6.2 Completed recording form for Monks Wood in August 1984.

seasonal and geographical variation in the flight period of the meadow brown *Maniola jurtina* and the hedge brown *Pyronia tithonus*. He described the flight period in terms of mean flight date and standard deviation about the mean date, and showed that in both species mean flight date was closely correlated with June temperature. These butterflies fly mainly in July and August; thus June temperature may be assumed to affect the rate of development of late larval and pupal stages. Surprisingly, in view of this relationship, there was no tendency in either species for the flight period to be earlier in the south of England than in the north. Brakefield thought that such a relationship might occur, but that it was not shown in the data he studied because he did not consider many sites in the north of England or Scotland. However Pollard (unpublished) showed that in the case of another satyrid, the grayling *Hipparchia semele*, in a range of sites covering over six degrees of latitude, showed no indication of later emergence in the north.

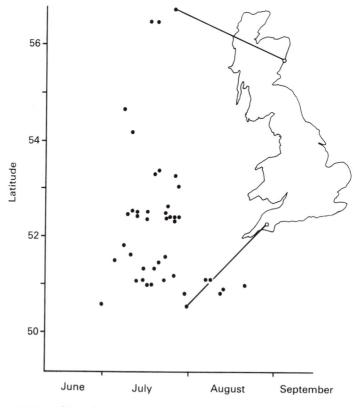

Figure 6.3 Mean flight date for the meadow brown (*Maniola jurtina*) at sites in the monitoring scheme in 1988.

The meadow brown appears to be unusual in that the length of the flight period is much greater in the south than in the north (Brakefield 1987). There is also much greater variability in the mean flight date in the south (Figure 6.3), and the species may be constrained to fly in the central part of the summer in the north.

The phenological data from the monitoring scheme are as yet largely unexplored. Apart from Brakefield's study, Dennis (1985) has studied the voltinism of the small tortoiseshell in different seasons and in different parts of the country. Pollard and Hall (1980) have observed that the flight period of the brimstone *Gonepteryx rhamni* differed at woodland and wetland sites in eastern England. They inferred from this that there may be movement of brimstones between the two types of site, with overwintering mainly in the woodlands and breeding mainly in the wetlands. As is usually the case, data from monitoring may indicate features of interest, but more detailed research is required to confirm or deny them.

(b) Migration

A few of the British butterflies do not usually overwinter in this country, but immigrate each year. The nature of these migrations is the subject of some debate. For example, Baker (1972) has suggested that the migration of the painted lady, *Cynthia cardai* is essentially short-range, with individuals seen in spring or early summer overwintering in the south and west of Britain or in the near continent. A more general view is that it migrates more or less erratically on a continental scale. The pattern of occurrence, both seasonal and geographical, at sites in the monitoring scheme should provide evidence in favour of one or other of these views. In fact, the pattern of occurrence of the painted lady varies very greatly from year to year, with occasional major movements at various times during the summer (e.g. Pollard 1982a), suggesting that migration is not of the regular short-range type envisaged by Baker. In contrast, the pattern of occurrence of the other common migrant, the red admiral *Vanessa atalanta*, is much more regular and more consistent with Baker's views.

Table 6.1 Clouded yellow butterflies at sites in the monitoring scheme 1976–88

	1976	77	78	79	80	81	82	83	84	85	86	87	88
No. of sites at which seen	1	1	0	2	4	6	5	48	9	1	1	2	0
Sum of index values	1	3	0	2	21	16	19	601	33	1	2	3	0

The clouded yellow *Colias croceus* is a rarer migrant than the two vanessids. During the period of monitoring, the clouded yellow was only abundant in 1983. In the years prior to 1983 and in the following year there were moderate numbers, but in all other years there were very few (Table 6.1). It is not known whether this pattern is due to chance or whether this migrant tends to build up numbers over several years before an 'outbreak'. Many years of monitoring are required before the significance of such patterns can be judged.

(c) Collated trends

The overall fluctuations in abundance, as shown by the collated trends, show widely different patterns for different species. Data for six species (Table 6.2) are used to illustrate the variability. The orange tip *Anthocaris cardamines* data have been amongst the most stable, with a range of index values of less than twofold over the recording period. In sharp contrast, the migrant painted lady has very large fluctuations from year to year and an overall range of × 65. Such large fluctuations are not surprising in a migrant species whose abundance in a particular year depends both on the numbers that immigrate and on breeding success within this country. The ringlet *Aphantopus hyperantus* has fairly small year-to-year changes, apart from a sharp fall in numbers from 1976 to 1977. However, there was a steady rise in index values from 1977 to the mid 1980s and an overall range of × 12. In contrast, the grizzled skipper *Pyrgus malvae*, one of the least common of the species for which index values are produced, has declined fairly steadily over the recording period. The data for both species in Table 6.2 which have two generations a year fluctuate considerably. The two generations of the wall *Lasiommata megera* have a regular pattern, with the second generation index always larger than the first. The holly blue *Celastrina argiolus* has no such pattern; there have been very large fluctuations in numbers, as have long been considered characteristic of this species (e.g. Frohawk 1934).

Although the collated index values should not be treated as national index values, it seems possible that for some species they may be a good approximation to national values. These are the very mobile species such as the large white *Pieris brassicae*, small white *Pieris rapae*, peacock *Inachis io*, small tortoiseshell *Aglais urticae* and the migrant species. All of these species occur on virtually all sites, but by far the most important breeding area for most of them is commercial farmland. The pierid species are pests of cultivated brassicas and several of the others feed as larvae on nettles along field edges. These butterflies are often seen, feeding at flowers or simply in flight, at sites where their food-plants are absent. Trends in abundance are likely to depend on farming practices and these trends will probably be reflected quite closely in the numbers seen away from farmland, such as on the many nature reserves in the monitoring scheme.

Table 6.2 All sites index values for six selected butterfly species 1976–88

	1976	77	78	79	80	81	82	83	84	85	86	87	88
Grizzled skipper	100	86	40	25	40	14	17	21	29	34	21	22	14
Orange tip	100	84	105	102	117	98	78	106	136	126	85	132	76
Painted lady	100	95	18	76	1170	34	855	232	21	387	89	48	741
Ringlet	100	33	59	102	146	168	325	317	404	301	358	372	264
Holly blue													
1st gen.	13	37	39	27	21	4	13	16	55	96	14	5	4
2nd gen.	100	56	18	41	9	3	14	35	90	37	4	1	4
Wall	50	7	9	12	15	19	42	39	28	13	8	10	13
	100	20	38	41	37	69	113	109	74	19	22	23	27

Most species have shown considerable fluctuations in numbers over the recording period. Such fluctuations make any underlying trends there may be, difficult to detect. Nevertheless, simple regression analysis of the data for the 29 species for which collated index values are produced shows six significant trends, five increases and one decline. The large skipper, peacock, comma, speckled wood and ringlet index values have increased significantly, while those of the grizzled skipper have declined. This analysis should be treated with caution, both because the sites are not representative and because the collation of data is not without problems (Chapters 2 and 12). However, it is of interest that the one species to decline, the grizzled skipper, is a local species which in the monitoring scheme and perhaps elsewhere is largely restricted to nature reserves. In contrast, the species which have increased are relatively common and include three which have expanded their ranges in the recording period.

One of the most conspicuous features of the data has been the synchrony of fluctuations over large areas (Pollard *et al.*, 1986). This synchrony applies not only to butterflies which range widely over large areas of countryside, but also to the many species which have largely discrete local populations (Thomas 1984). Clearly, these local populations are influenced by factors which affect wide areas and, almost certainly, variation in weather from year to year is most important.

Correlations between broad measures of temperature, rainfall and butterfly index values have been examined (Pollard 1988). It is very difficult to separate correlations caused by real effects of weather on index values from those that are simply due to chance, especially in such a short run of data (13 years of recording is a long period relative to most population studies of animals, but very short for correlation analysis). However, a strong association between increase in numbers and early summer temperatures may well be causal. The phenological data suggested that in warm years emergence is early because of rapid larval and pupal development. In such years it is also likely that survival of the immature stages is good and numbers will rise. As is always the case, monitoring cannot do more than suggest such possibilities, more detailed population studies are needed to establish the true causes of population changes. Relationships with temperature are not likely to be simple; another correlation was between warm dry summers and a decline in numbers the following year. The ringlet data for 1976–7 (Table 6.2), following the severe 1976 drought, provide one such example and there are several others. These aspects of monitoring of butterflies are potentially of special importance if expected climatic changes occur. As both temperature and rainfall are likely to have effects on butterfly populations, responses to climatic change may well be complex.

(d) Introduction, colonisation and extinction

It is now quite a common procedure to introduce rare butterflies to nature reserves or other sites where they once occurred, but have become extinct. Lessons can be learnt from both the success and failure of such introductions. One example in the monitoring scheme is the introduction of the Adonis blue *Lysandra bellargus* to the chalk downland nature reserve Old Winchester

Table 6.3 Index values of the Adonis blue at the few sites in the monitoring scheme for which several years' data are available. The Adonis blue was introduced to Old Winchester Hill in 1981, as indicated by an asterisk

	1976	77	78	79	80	81	82	83	84	85	86	87	88
Swanage													
1st gen.	188	13	35	99	237	101	384	168	268	138	91		
2nd gen.	17	17	67	254	493	271	1247	1006	699	76	334		
Castle Hill													
1st gen.			12	1	44	10	22	70	650	166	53	65	22
2nd gen.			5	36	14	20	179	419	483	30	70	31	15
Martin Down													
1st gen.			1	0	0	0	3	6	24	5	4	0	
2nd gen.			0	0	0	4	16	21	11	8	6	10	
Old Winchester Hill													
1st gen.	0	0	0	0	0	2*	10	20	38	6	0	3	3
2nd gen.	0	0	0	0	0	14	75	115	103	11	17	2	3

Hill (Table 6.3). This species prospered initially, but is now in low numbers. The decline may be related to weather rather than site factors, as the Adonis blue is in low numbers at the few other sites in the monitoring scheme where it is present.

In the relatively short period that the monitoring scheme has been in existence, several butterflies have expanded their ranges (information from various sources). These species include the comma, ringlet and speckled wood, which have increased significantly in numbers and also the small skipper, hedge brown and orange tip. The monitoring scheme was not established to detect such changes in range and in some cases there is no site in the area of expansion; in other cases, however, several sites in the scheme have apparently been colonised. It is impossible to be sure that a species has indeed recently colonised a site, especially when numbers seen are small. However, in a few cases, a species has shown a rapid increase in the first year or two after the initial sighting and subsequently has fluctuated more or less in line with wider trends. This is good evidence that colonisation has occurred and an illustration of the value of synoptic recording.

(e) *Site studies*

Butterfly monitoring provides information on the distribution of butterflies within a site, as the recording routes are divided into sections with separate counts for each. However, the interpretation of this information varies very greatly for different species. Some butteflies do not usually fly far and the distribution of counts may give a good indication of the breeding areas; an example of such a species, discussed in the case study, is the dingy skipper *Erynnis tages*. Other butterflies range widely over the countryside and have populations which are difficult or impossible to define. Extreme examples of these species are the migrants, such as the painted lady which recolonises Britain each year, probably from the Mediterranean. In such cases, the distribution along a route may simply reflect the occurrence of flowers favoured for adult feeding. Between these extremes, there are species in which the counts may be concentrated in the breeding areas, but nevertheless they frequently fly into adjoining areas. Unfortunately, for many butterflies, the extent of movement is not known in any detail and this lack of knowledge limits the value of much of the distributional data in the scheme. Thomas (1984) has summarised the available information on the mobility of butterflies, categorising the species into those with closed and open populations.

One of the main aims of the scheme is to assess the impact of factors such as habitat change on the butterflies of individual sites. Such assessments are made by comparisons of individual site data with collated trends and by changes in the relative numbers of butterflies seen in different sections of the recording route. One such study, at a woodland site, has been published

(Pollard 1982b), while Pollard *et al.* (1986) give summary accounts of 86 sites. In this chapter, the problems and potential for this type of study are shown by means of a case study of a chalk downland site.

6.3 CASE STUDY: CASTLE HILL NATIONAL NATURE RESERVE

With the collaboration of Roy Leverton

Castle Hill National Nature Reserve (Figure 6.4) is a 45 ha area of chalk downland slopes and dry valleys in the South Downs near Lewes. The basic aspect is southerly, but there are east- and west-facing slopes. The flatter tops of the slopes are capped with gorse and there is also elder, hawthorn and abundant bramble. The remainder of the reserve is grassland of two main types, herb-rich old grassland on the steeper slopes and simpler grassland on areas which were arable in the 1960s. The reserve is largely surrounded by arable farmland and improved grassland, but there is also some herb-rich grassland outside the reserve.

The area was declared a nature reserve in 1975 and butterfly monitoring began in 1978. Apart from characteristic chalk-downland plants and butterflies, the reserve has many other features of interest, including scarce species of orchids and the rare bush cricket *Decticus verrucivorous*.

Management of Castle Hill is primarily by grazing, formerly by cattle alone, but from 1980 to 1986 the management differed on the two halves of the reserve. The western half, Newmarket Hill, continued to be grazed by cattle, mainly in winter. The eastern half, Castle Hill itself, was managed by sheep alone from 1980–86 and this grazing was mainly in summer. A problem in the development of the grassland has been the spread of the tussock-forming tor grass *Brachypodium pinnatum* on the herb-rich downland. There are no long-term botanical data, but there is no doubt that the herbs and finer grasses characteristic of chalk downs have declined in abundance where tor grass has become established. The spread of tor grass has been much more rapid on the sheep-grazed eastern area (warden's report, Nature Conservancy Council internal document), apparently because the sheep graze the turf around the tor grass tussocks but not the tussocks themselves. There has been a limited amount of mowing in this eastern area, to help with the control of tor grass. In contrast, cattle grazing in the western half of the reserve appears to have maintained a herb-rich sward much more successfully. In 1980, a combination of drought and a high density of sheep led to severe over-grazing. In subsequent years lower densities of sheep were used and there were no similar problems. Since 1987, the management of the reserve has no longer been consistent to the two halves; this study is therefore limited to the 1978–87 period. In addition to the grazing by sheep and cattle,

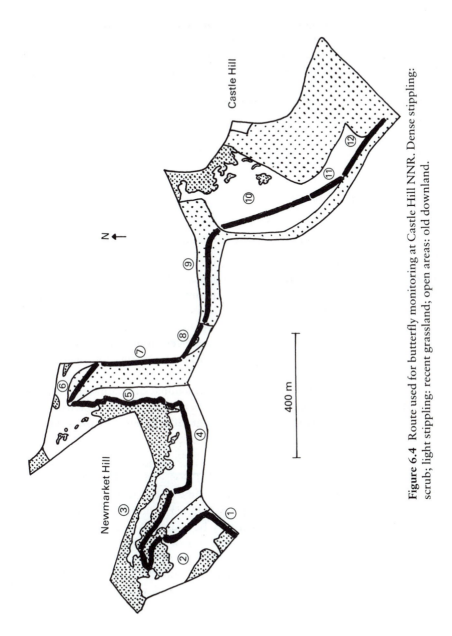

Figure 6.4 Route used for butterfly monitoring at Castle Hill NNR. Dense stippling: scrub; light stippling: recent grassland; open areas: old downland.

rabbit grazing is widespread and variable but has not obscured the sharp differences between the eastern and western halves.

The route used for butterfly monitoring (Figure 6.4) is largely restricted to grassland areas. One section is through scrub, eight are through old downland and two through grassland on recently abandoned arable. The remaining section (no. 7) follows the junction between grassland on abandoned arable within the reserve and rich old downland turf outside the reserve. Sections 1–8 are in the western cattle-grazed area and 9–12 in the eastern sheep-grazed area. Section 12 differs from the rest of the sheep-grazed area in that sheep have been excluded in summer since the heavy grazing of 1980.

6.3.1 Methods

The detection of local effects of habitat change makes use of comparisons of local data with trends at regional or national levels. This method is based on the following assumptions.

1. Annual fluctuations at regional and national levels are caused mainly by variation in weather conditions, sometimes acting directly, sometimes indirectly through effects on food-plants, predators, parasites or other factors. A major effect of weather on fluctuations is indicated by the similarity of fluctuations in index values over large areas (Pollard *et al.* 1986) and by correlations between weather conditions and fluctuations (Pollard 1988).
2. Departures from these wider trends, at individual sites, are frequently the result of local habitat change; departures may be sudden, due to drastic management or other radical change, or gradual, often due to successional changes in vegetation.

In previous studies, departures from wider trends have been assessed subjectively. Here, an objective test is used. The size of the departure (in logs) from the collated trend is regressed against time (years) using standard linear regression. A significant slope ($p < 0.05$) is taken as indicating a site-specific departure from wider trends, which may be caused by local changes in the biotope. The method is most appropriate for detection of gradual changes and cannot be used for the detection of changes due to immediate effects of, say, the felling of woodland trees. Similarly, in the chalk grassland example described here, short-term effects of summer grazing would not be detected.

6.3.2 Results

A total of 26 butterfly species have been recorded during the years 1978–87 (Table 6.4). Of these, 16 species are thought to have more or less discrete

Table 6.4 Index values of butterflies at Castle Hill 1978–87. Where there are two distinct flight periods these are given on separate lines

	1978	79	80	81	82	83	84	85	86	87
Small/Essex skipper	399	643	374	718	518	521	836	710	1134	234
Large skipper	29	36	52	23	56	69	120	101	156	111
Dingy skipper	50	74	22	37	53	60	139	116	84	113
Clouded yellow†	0	0	0	3	3	28	0	0	0	0
Brimstone	0	0	0	0	0	0	0	0	2	0
	0	0	0	0	2	1	1	0	0	0
Large white	6	20	5	5	10	9	1	6	8	0
	48	22	3	2	16	16	2	8	29	8
Small white	26	10	7	18	8	7	15	8	6	4
	149	33	23	32	190	116	82	49	188	73
Green-veined white	1	1	0	0	3	0	0	0	1	0
	16	25	7	1	19	2	2	2	11	4
Small copper	13	5	15	9	4	11	15	7	12	22
	35	24	40	24	210	62	52	49	71	100
Small blue	2	18	4	0	0	1	4	13	48	7
Brown argus	1	3	0	0	1	4	5	1	0	0
	4	8	5	2	30	44	42	6	9	8
Common blue	54	206	571	254	43	154	271	86	88	59
	130	431	177	145	440	835	415	58	143	161
Chalkhill blue	1835	2720	1240	585	1085	1530	3016	2540	4514	2790
Adonis blue	12	1	44	10	22	70	650	166	53	65
	5	36	14	20	179	419	483	30	70	31
Red admiral†	3	0	3	15	7	7	2	5	2	25
Painted lady†	0	3	5	3	34	3	0	13	7	1
Small tortoiseshell†	85	97	73	116	226	50	130	177	48	486
Peacock	*	17	16	10	12	22	28	32	9	8
	23	10	7	7	12	15	36	30	13	15
Comma†	0	7	3	6	5	9	2	6	3	6
Dark green fritillary	10	5	3	0	0	0	4	1	5	4
Speckled wood†	1	0	0	1	1	3	8	0	0	1
Wall	4	6	12	23	26	17	10	2	2	1
	12	14	30	59	150	48	23	0	14	2
Marbled white	52	40	17	62	101	297	482	683	1248	471
Hedge brown	2	12	7	3	14	23	41	17	18	6
Meadow brown	4922	2960	1467	2473	1792	1252	2704	1816	2960	1636
Small heath†	532	274	182	154	522	491	720	250	377	149

† Indicates species with an annual index which is the sum of two or more flight periods which are not discrete.

* Indicates insufficient data for the calculation of an index.

populations breeding within the reserve and in its immediate vicinity. The data for the two *Thymelicus* species (small and Essex skippers) are combined.

The index values of some species, such as the Adonis blue *Lysandra bellargus* and the marbled white *Melanargia galathea*, have fluctuated quite widely over the ten year period. However, in most cases, the butterflies have maintained similar index values relative to other species. The correlation coefficient calculated using the index values (log. transformation) of the 16 sedentary species in 1978 and in each subsequent year varies between 0.94 (1979) and 0.81 (1983) (in all cases, $p < 0.01$). There is some tendency for the successive correlations with the 1978 index values to decline over the period. This decline, tested by linear regression, was significant at $p < 0.05$.

6.3.3 Comparison with wider trends

Trends in index values at Castle Hill are compared with trends shown by sites in the south and south-east of England (i.e. within these regions of the Nature Conservancy Council). Comparisons with regional trends are possible for the 11 most abundant of the 16 sedentary species (Figure 6.5). Three species, the large skipper *Ochlodes venata*, marbled white and the hedge brown show significant relative increase (departures from regional trends) over the recording period. These species are all associated with tall grassland, or grassland with scrub (Table 6.5). One species, the common blue *Polyommatus icarus*, showed a small, but significant relative decline. A feature of the Castle Hill data, even for those species which depart from regional trends, is the similarity in the pattern of fluctuations at Castle Hill to those of the region.

Of the five species for which comparisons with wider trends are not possible, four are sufficiently rare to be of special conservation interest. The fluctuations of the Adonis blue resemble those at the few other sites in the monitoring scheme where it is present (Table 6.3), as also do those of the small blue *Cupido minimus* and brown argus *Aricia agestis*. The dark green fritillary *Argynnis aglaia* is very rare at Castle Hill and little can be said about changes in its numbers. The speckled wood is also rare on the transect route, although it is a common woodland butterfly; it has been recorded almost exclusively in the scrub-dominated section 2.

6.3.4 Section data

The index values for each section of the recording route (Figure 6.6) show that, although the distribution of different species varies considerably, in some sections large numbers of several species have been recorded. Of the four richest sections, 1, 4 and 12 are old downland, while section 7 is thought to attract butterflies from the downland outside the reserve which

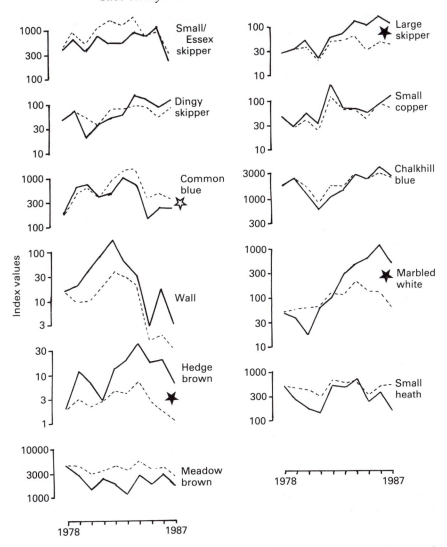

Figure 6.5 Comparison between trends in index values at Castle Hill (solid line) and in collated values for sites in southern and south-east England (dotted line). Stars indicate significant differences at p < 0.05.

immediately adjoins this section (Figure 6.4). In some cases, species with very different habitat requirements, including those that prefer different sward heights (Table 6.5), were abundant in the same sections. For example, the Adonis blue and marbled white, which require very short turf and long turf respectively, both had high counts in sections 4 and 12, although there were

Table 6.5 Preferred sward heights and food plants of the main butterfly species considered in the Castle Hill study. Sward height information simplified from B.U.T.T. (1986); food plant based on Heath *et al.* (1984). The small and Essex skippers are not separated in this study

Sward height (cm)	Major food plants
<1	
Adonis blue	Horseshoe vetch
1–5	
Dingy skipper	Birds-foot trefoil
Chalkhill blue	Horseshoe vetch
Small heath	Various, mainly fine-leaved grasses
1–10	
Small copper	Sorrels
Common blue	Birds-foot trefoil, black medick
Wall	Tor grass and various other grasses
Meadow brown	Various, mainly fine-leaved grasses
5–15	
Large skipper	Cock's foot and some other grasses
Marbled white	Sheeps fescue, tor grass and various other grasses
Hedge brown	Various, mainly fine-leaved, grasses
>15	
Small and Essex skippers	Flowering stems of some grasses

other sections, such as 5, in which their counts differed markedly. Sections 6 and 9, through grassland on former arable areas, had low numbers of several species, notably the dingy skipper and chalkhill blue *Lysandra coridon*. All species were poorly represented in the scrub-dominated section 2 except the large skipper and speckled wood.

Index values for each of the 12 sections of the transect route at Castle Hill were compared with regional trends in the same way as was done for the complete site index values (Figure 6.6). Significant departure from regional trends gave little indication of major differences in trends in different areas of

Figure 6.6 Percentage of total index values in each section at Castle Hill (mean and range). Dense stippling = scrub: light stippling = new grassland on former arable areas: open = old downland. Significant departures from regional trends (p < 0.05) indicated by stars: solid stars = relative increases: open stars = relative declines. No regional comparisons available for the Adonis blue.

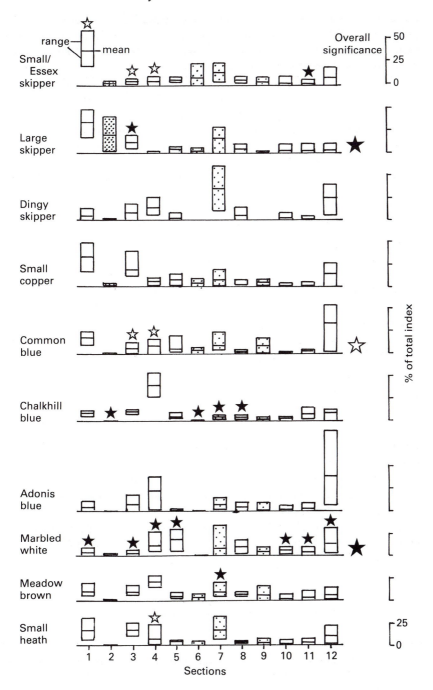

the reserve. Only the *Thymelicus* species showed departures in different directions in different parts of the reserve, with increases in the sheep-grazed eastern half and decreases in the cattle-grazed western half. The marbled white increased significantly in many sections in both halves of the reserve. The greatest increases, as judged by the steepness of the slope b, were in sections 10–12. The chalkhill blue increased significantly in four sections, but these had relatively low counts and did not contribute greatly to the overall site index, which did not itself depart significantly from regional trends.

The impact on butterflies of the period of heavy grazing by sheep in 1980 is illustrated by comparing the summed data for the affected sections 9–12 with those for sections 1–8. Very few butterflies were seen in these sections in 1980 and the index values fell sharply below those expected on the basis of changes elsewhere (Figure 6.7). Index values for other species declined similarly. In most cases, the 1981 index values appeared to have recovered.

6.3.5 Discussion

In general, species with high or low index levels at the start of the recording period retained similar levels over the ten years. The relative values of the

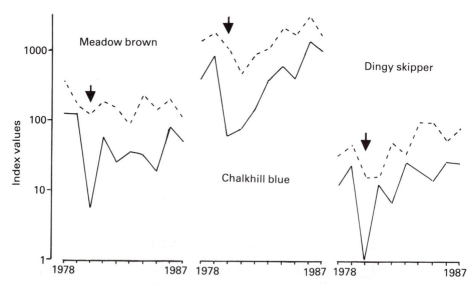

Figure 6.7 Trends in index values in an area of Castle Hill heavily grazed by sheep in 1980 (solid line) compared with trends in the sections not grazed by sheep (dotted line). A sharp decline in index values in the sheep-grazed area in 1980 (indicated by the arrows) was followed by a rapid recovery. Meadow brown index values divided by ten for convenience of presentation.

indices of abundance of different butterfly species are not measures of relative abundance, because the likelihood of seeing an individual varies from species to species. However, most species with low index values at Castle Hill have high values at some other sites and vice versa (Pollard *et al.* 1986), therefore it seems reasonable to suggest that relative abundances have remained rather stable. This stability in turn suggests that there have been no gross changes in the types of habitat available to butterflies at Castle Hill.

Comparison of data from Castle Hill with all sites in south and south-east England shows that the main features of both annual fluctuations and long-term trends at Castle Hill are very similar to changes over a much larger area. This indicates that the changes in abundance at Castle Hill, over the ten-year period, are largely due to factors which affect southern England as a whole. It is likely that variation in weather conditions plays an important role. Similarly the small, but significant, decline in the correlation of the index values in successive years with those in the first year seems to be caused largely by diverging regional trends in butterfly abundance.

However, three species, the large skipper, marbled white and hedge brown increased relative to regional trends. These are all species associated with long grassland or with the margins of scrub and grassland. This is in accord with the spread of the coarse, tussock-forming tor grass and a possible small expansion of brambles at the scrub margins. The most striking change was by the marbled white, which is strongly associated with long (5–15 cm) grassland (Table 6.5) on lightly grazed or ungrazed downs (Thomas 1986).

A relative increase of species associated with tall grassland might be expected to be accompanied by a concomitant decrease in species associated with short turf. This does not seem to have occurred. Of the species for which wider comparisons are possible, only the common blue declined significantly relative to the regional trend and this decline has been quite small. No comparison is possible for the Adonis blue, the species which is probably most dependent on short grassland, but the pattern of change of this species is broadly similar to that at other sites in the monitoring scheme.

It is perhaps surprising that the marbled white and Adonis blue, which have very different requirements in terms of sward height, tended to have high index values in the same sections. This apparent paradox may be because the swards are not uniform; a mosaic of short turf between tussocks of tor grass may provide good conditions for both species. Unfortunately, such a mosaic is likely to be a stage in the development of tor grass and it is difficult to maintain a sward in this condition.

Although the two parts of Castle Hill National Nature Reserve have been managed quite differently and this has affected the structure of the vegetation, the section data on butterfly counts show little clear response to the differences. Only the combined data for the small and Essex skippers have significant relative changes in opposite directions in the two parts, with decreases in the cattle-grazed area and increases in the sheep-grazed area. As the species may have responded differently, no simple interpretation of this result can be made.

It is possible that there have been changes in breeding success in different areas of the reserve, but that these changes have been obscured by movement of adult butterflies away from the areas where they emerge. The reasons for such movement may include separation of breeding areas from nectar sources and roosting areas, and may also be influenced by factors such as wind direction and strength.

The sharp reduction in counts in the heavily sheep-grazed areas of the reserve in 1980 was followed by a recovery in 1981. This suggests that most species were sufficiently mobile to recolonise these areas quickly when they again became suitable for butterflies. In general, the index values in the sheep-grazed area, such as those shown in Figure 6.7 did not follow regional trends (Figure 6.6) as closely as did those in the cattle-grazed area. The sheep-grazing, which was in summer, was variable in timing and intensity from year to year and this is likely to have caused considerable variation in the structure of the sward. Winter cattle-grazing would have much less impact.

Relatively low numbers of butterflies were recorded in grassland recently established on arable land (sections 6 and 9). The only species to increase significantly in either of these sections was the chalkhill blue in section 6, but the food-plant, horseshoe vetch *Hippocrepis comosa*, is virtually absent from these sections. The trend must be related to movement of butterflies in to the area, presumably to feed at flowers.

In summary, the results of butterfly monitoring at Castle Hill provide evidence that butterflies associated with long grassland have increased there more than expected on the basis of regional trends. However, there is no evidence of relative decline of the species which require a shorter turf, although, for some of the rarer species, comparative data are not available. On present evidence, it seems likely that the changes in sward structure and composition, notably the spread of tor grass, are having some impact on the butterflies at Castle Hill, but the situation is not critical and there is still time to experiment with management techniques.

6.4 LIMITATIONS AND POTENTIAL OF BUTTERFLY MONITORING

The main aims of butterfly monitoring, outlined at the beginning of this chapter were, in brief: to provide information at both national level and individual site level on changes in butterfly numbers and to interpret the site data in terms of habitat change. In this section the extent to which these aims have been realised will be assessed.

The recording method seems robust and combines many of the features required for monitoring; that is, it is simple, quick and can be used in a wide range of biotopes and under a range of weather conditions. The main weakness may be differences between individuals in their recording, but this is overcome to some extent because individual recorders have continued for many years at the same sites.

A major limitation to the 'national' data lies in the selection of sites. Ideally, the sites in a monitoring scheme should be selected, by an objective method, to be representative of the different types of countryside in Britain. In practice, because resources are limited, the sites include a predominance of nature reserves. It has been argued here that the results for some species may be close to true national data, but the extent to which this is true is not known and it is advisable to refer to all collated data as 'all site' data.

In addition to these difficulties of site selection, the selection of the recording route within the site also presents problems. It is likely that a recorder will include the best butterfly areas in his route. If there is no management, such a favoured area may deteriorate as succession of vegetation progresses; if there is special management for butterflies, it is likely to be concentrated along the recording route and counts may tend to increase. It is in the nature of butterfly monitoring, that the very fact that an area is being monitored has an impact on the site and so on the abundance of butterflies. The effects of such biases cannot be easily assessed. It is possible that these are largely theoretical problems which are of little importance in practice. It is certainly likely that major changes, such as may be expected in the event of climatic change, will overwhelm any such effects.

Trends in the collated index values are of considerable interest, whether or not they show true national trends. Although some significant trends have been observed over the 13-year recording period, the extent of short-term fluctuations makes the detection of trends difficult. For some species, longer term trends may only be evident after several periods of abundance and scarcity. Similarly, it requires many years to establish whether these periods of abundance and scarcity are erratic, related to external factors, such as weather, or are more or less regular, related to some intrinsic cycling in the populations of a species.

The assessment of local impacts on the abundance of butterflies at individual sites depends in part on comparisons with trends at other sites, as in the case study at Castle Hill. The question to be asked is whether butterflies at a particular site are faring better or worse than at other sites in the scheme, as shown by comparison with the collated results. If the possible biases in the data do indeed have important effects then the site assessments could be misleading. For example, a particular form of woodland management might be detrimental, but the species of interest could, in theory, show an increase relative to other sites in the scheme. This could happen if increased shading of woodlands in general caused a greater decline in the collated index than that observed at the individual site. This method of site assessment has considerable value, but the possibility of misleading results should not be forgotten. Conclusions based on comparisons with other sites may be made with more confidence if they are reinforced by comparisons between different areas within a site. Here, however the general lack of knowledge of the population structure of butterflies may make interpretation difficult, as in the case study. There is a clear need for further research on the dispersion of butterflies in relation to their breeding areas.

A major limitation in site assessments is the lack of information on vegetation changes. For many sites, there are quite detailed data on changes in butterfly numbers but little or none on the structure and composition of vegetation. Given the large amount of time required for botanical monitoring, this lack of data is not surprising. There is, however, a clear case for combining botanical and butterfly monitoring (and perhaps that of other organisms) at a limited number of 'core' sites. Such comprehensive recording should do much to clarify the effects of changes in vegetation on insect populations.

In spite of the limitations discussed, the Butterfly Monitoring Scheme has shown the value of simple recording conducted systematically over a long period. Synoptic recording certainly aids the interpretation of results from individual sites. In addition to monitoring changes in numbers, the data from the scheme on the phenology of butterflies, on migration and on local colonisation and extinction are of considerable interest. Although monitoring should be started with clear aims, it is quite likely that, as was the case here, some of the most interesting aspects of the results will be unexpected. In this case, one of the minor aims was to examine the relationship between weather and butterfly numbers. If the scheme continues over the next decades, this aspect may prove one of the most important and interesting.

ACKNOWLEDGEMENTS

I would like to thank the Nature Conservancy Council and the Institute of Terrestrial Ecology (N.E.R.C), who have supported the Butterfly Monitoring Scheme since its inception. The scheme depends on the enthusiasm of the recorders who count the butterflies and special thanks are due to the many who have taken part. Alan Bowley, the warden of Castle Hill National Nature Reserve, provided invaluable help with the case study.

REFERENCES

Baker, R.R. (1972) The geographical origin of the British spring individuals of the butterflies *Vanessa atalanta* (L.) and *Vanessa cardui* (L.), *Journal of Entomology* A **46**, 185–96.

Brakefield, P.M. (1987) Geographical variation in, and temperature effects on, the phenology of *Maniola jurtina* and *Pyronia tithonus* (Lepidoptera, Satyrinae) in England and Wales, *Ecological Entomology*, **12**, 139–48.

B.U.T.T. (Butterflies Under Threat Team) (1986) *The Management of Chalk Grassland for Butterflies*, Nature Conservancy Council, Peterborough.

Cochran, W.G. (1963) *Sampling techniques* (2nd ed.), Wiley, New York.

Dennis, R.L.H. (1985) Voltinism in British *Aglais urticae* (L.) (Lep. Nymphalidae): variation in space and time, *British Entomological and Natural History Society, Proceedings and Transactions*, **18**, 51–61.

Frohawk, F.W. (1934) *The Complete Book of British Butterflies*, Ward Lock, London.

Heath, J., Pollard, E. and Thomas, J.A. (1984) *Atlas of Butterflies in Britain and Ireland*, Viking, London.

Pollard, E. (1982a) Observations on the migrating behaviour of the painted lady butterfly *Vanessa cardui* (L.) (Lepidoptera Nymphalidae), *Entomologist's Gazette*, **33**, 99–103.

Pollard, E. (1982b) Monitoring the abundance of butterflies in relation to the management of a nature reserve, *Biological Conservation*, **24**, 317–28.

Pollard, E. (1988) Temperature, rainfall and butterfly numbers, *Journal of Applied Ecology*, **25**, 819–28.

Pollard, E. and Hall, M.L. (1980) Possible movement of *Gonepteryx rhamni* between hibernating and breeding areas, *Entomologist's Gazette*, **31**, 217–20.

Pollard, E., Hall, M.L and Bibby, T.J. (1986) *Monitoring the Abundance of Butterflies 1976–85*, Nature Conservancy Council, Peterborough.

Thomas, J.A. (1984) The conservation of butterflies in temperate countries: past efforts and lessons for the future, in *The Biology of Butterflies* (eds R.I. Vane-Wright and P.R. Ackery) (Symposium of the Royal Entomological Society No. 11), Academic Press, London.

Thomas, J.A. (1986) *Butterflies of the British Isles*, Newnes, London.

— 7 —————————————————

Monitoring terrestrial breeding bird populations

STEPHEN R. BAILLIE

7.1 INTRODUCTION

It is important to monitor bird populations for two main reasons. First, birds form an important and popular wildlife resource which is worth conserving in its own right. Second, given that it is impractical to monitor all groups of organisms on a wide scale, birds provide valuable indicators of the state of the environment. They are usually high in food chains and so particularly susceptible to environmental changes. This is well illustrated by the population declines and breeding failures of raptors which drew attention to the harmful effects of persistent organochlorine pesticides (Newton 1979, 1986; Ratcliffe 1980).

Although bird population monitoring is not without difficulty, it is easier than for many other groups of animals. Much has been said earlier in this book about the needs to define clear objectives for monitoring schemes and to establish standards against which changes can be measured. Extensive long-term data are available for bird populations, providing a description of past population fluctuations with which recent changes can be compared. In addition, it is often possible to measure breeding success and survival rates of birds as well as population size, which assists in the interpretation of population changes and may even draw attention to problems before populations start to decline.

Throughout much of Europe and North America, many amateur bird-watchers contribute enormous amounts of time, effort and money to monitoring schemes. Thus it is possible to operate extremely cost-effective monitoring schemes for birds. Most amateurs participate in monitoring schemes because they find it an enjoyable hobby. Schemes which involve complex procedures or paperwork, or which require observers to spend large amounts of time in areas with few birds, are inevitably unpopular. Optimal monitoring schemes for birds must often balance the desirability of a statistically rigorous study design (giving unbiased estimates which are easy to interpret) against the likely number of observers (more observers giving increased precision). Most terrestrial bird monitoring schemes in Britain and Ireland are organised by the British Trust for Ornithology (BTO) which co-ordinates the work of amateur bird-watchers. Much of this work is carried out under contract to the Nature Conservancy Council.

This chapter provides a brief overview of the main methods used for monitoring breeding populations of terrestrial birds, including their reproductive success and survival. The main BTO schemes that monitor terrestrial birds in Britain and Ireland are outlined, together with the new Integrated Population Monitoring Programme, which will combine the results from different schemes to provide a more thorough assessment of population performance. Limitations of present monitoring schemes and priorities for future developments are briefly discussed.

7.2 METHODS FOR MONITORING BIRD POPULATIONS

7.2.1 Territory mapping

This technique, known in North America as spot mapping, is widely used for estimating numbers of pairs of terrestrial passerines during the breeding season. International standards for fieldwork and data analysis have been agreed (Anon 1969), but various modifications and enhancements of the technique have been developed for specific studies (Falls 1981; Wiens 1969).

Standard territory mapping involves a series of visits to the study plot, spread throughout the breeding season. On each visit the locations of all birds seen or heard are recorded on a large-scale map (Figure 7.1a). Species, sex, song and other activities are recorded on the maps, with lines joining records of the same bird and broken lines indicating simultaneous registrations of different birds. At the end of the season this information is transferred onto maps that contain all the information for each species, using different letters to indicate different visits. Clusters of registrations are delineated on the maps and each cluster is taken to represent one territory (Figure 7.1b). Territory clusters are usually delineated manually; although computer algorithms for such analyses have been devised (North 1977, 1979), they are not yet able to use all the information available to human analysts.

There have been many studies of factors that may bias results from territory mapping. The number of territory clusters depends on the detection probability of individual birds on each visit, the number of visits, and the minimum number of registrations (on different dates) required to form a cluster. Low detection probabilities, low numbers of visits, or high numbers of registrations required for a cluster lead to underestimation, while the converse leads to overestimation (Svensson 1979; Dawson 1985; Verner 1985). Results from territory mapping can differ between observers and between analysts (Best 1975; Enemar *et al.* 1978; O'Connor 1981), although

(a)

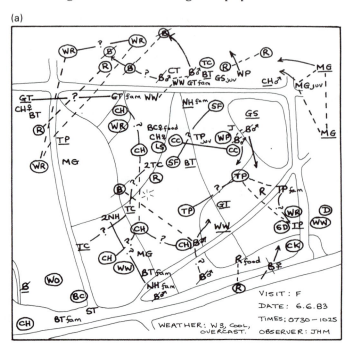

(b)

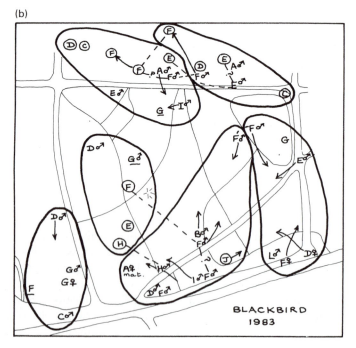

agreement between analysts can be improved by training (see Section 7.3.1). Territories at the edges of the plots provide difficulties of classification (Marchant 1981), which are particularly severe for species whose territories are large relative to plot size. Other limitations of territory mapping are that studies carried out on individual plots have no replication, so confidence limits cannot be calculated, and that the method takes no account of non-territorial individuals.

Despite these limitations, many workers consider that careful territory mapping is the most practical method for determining the numbers of territorial breeders present within a given area (Tomialojc 1987; Verner 1985; but see Dawson 1985). Reliable estimates can usually be obtained by mapping the territories of individually colour-marked birds (total mapping – Verner 1985) but the resources required for this approach prohibit its use for most studies. The few comparisons between standard territory mapping and colour-marking suggest that territory mapping does not always provide good measures of absolute density (Verner 1985).

7.2.2 Line transects

Line transects involve observers walking a fixed straight line route and recording the birds that they see on either side of the line. They can be used at any time of year to record all classes of individuals within the population and are best suited to large areas of continuous habitat through which observers can move without difficulty. Line transects have been used widely to study bird populations, particularly in Finland (Merikallio 1958; Jarvinen and Vaisanen 1976) and in North America, but I am not aware of their use in any national monitoring schemes for breeding birds.

Several different transect methods have been developed. Unlimited distance transects provide only relative estimates of numbers, while strip transects, fixed-distance line transects and variable-distance line transects

---

Figure 7.1 Examples of part of a visit and species map from a territory mapping census carried out within the woodland CBC scheme.

(a) Visit map. Two-letter codes represent different species, circles indicate singing birds, solid lines join records of the same individual and broken lines join records of different individuals. Blackbird registrations (B) have already been copied to the species map and cancelled with a light stroke of the pen.

(b) Blackbird species map from the same census as shown in Figure 7.1(a). On transfer to the species map the B for Blackbird has been replaced by the appropriate visit letter, F in the case of the registrations shown in Figure 7.1(a), but symbols indicating sex, song and movements have not been changed. The territory clusters have been delineated, and six territories found on this portion of the plot.

provide estimates of absolute densities (although these are often severely biased, Verner 1985). Annual monitoring usually requires comparisons of counts from the same time of year within species, in which case estimates of absolute densities are not required (Section 7.2.3).

The three transect methods providing measures of absolute densities all use information on the distance between the birds and the transect line to estimate the effective area being censused. Strip transects simply count all birds within a set distance from the centre line and assume that within this area all birds are detected. Birds seen outside the strip are not recorded. In fixed-distance line transects birds are recorded as being either within a central zone (usually 25 m either side of the line) or beyond it. Densities are calculated by assuming that the fall off in detectability with distance from the transect line follows a linear function (Jarvinen and Vaisanen 1975).

Variable-distance line transects involve recording either the perpendicular distance from each bird to the transect line or categorising individuals into distance zones (Emlen 1971, 1977). A statistical model is then fitted to the data to describe the fall off in detectability with increasing distance from the line (Burnham *et al.* 1980). Models assume either that all birds exactly on the transect line are detected (ungrouped data), or that all the birds in the first distance zone are detected (grouped data). The main difficulty in these analyses is that it is necessary to assume that birds do not move away from the transect line in response to an observer. However, such movement does inevitably occur, giving rise to unreliable estimates of densities. Where there is substantial movement in response to the observer, Burnham *et al.* (1980) recommend against the use of line transects.

7.2.3 Point counts

Point counts provide similar information to line transects, being suitable for use throughout the year and for counting all categories of individuals. They involve counting birds from fixed points during a specified time interval. As with line transects, unlimited distance counts can be used to compare relative abundance while methods involving distance estimation must be used where estimates of density are needed (see Section 7.2.4).

Estimation of densities from point count data is essentially a special case of the line transect problem, requiring both the assumption that all birds are detected at the centre of the count area, and the computation of a function describing the fall off in detectability with distance from the centre of the circle. As with line transects, both variable distance (Reynolds *et al.* 1980) and fixed distance methods (Buckland 1987) have been developed. The area sampled by a point count increases geometrically with distance from the centre of the circle, while with line transects area sampled increases linearly from the transect line. This means that, with point counts, small errors in detecting birds close to the observer can seriously bias density estimates (Verner 1985).

The duration of counts is an important aspect of study design. If the period is too short, individuals are likely to be missed, while if it is too long, densities are likely to be overestimated, because birds move through the count area or are counted twice. Periods in common use range from three to 20 minutes, the most common period for European monitoring schemes being five minutes. A study of count duration in six habitats in Britain suggested that they did not need to be longer than 10 minutes, and that five minutes was adequate for many purposes (Fuller and Langslow 1984).

7.2.4 Comparison of counting methods

All counting methods in current use have serious limitations which must be taken into account when selecting monitoring methods and interpreting their results. Monitoring usually involves between-year comparisons, for single species, of data gathered at the same time of year. For these it is reasonable to assume that detectability does not vary between years. Unlimited distance line transects or point counts are suitable for annual monitoring, as is territory mapping. Line transects are best suited to large areas of homogeneous habitat while point counts are more suitable where habitat is patchy, as in many parts of western Europe and North America. There is continuing debate as to whether territory mapping provides better monitoring information than point counts and line transects (Dawson 1981a, 1981b; Verner 1985). Territory mapping usually provides a more complete assessment of the birds present on particular study plots, but because it is very labour intensive it limits the areas and habitats that can be sampled. There have been no studies to compare the precision and cost of point counts and territory mapping at the scale of regional or national monitoring programmes.

Outside the breeding season, territory mapping is unsuitable and line transects or point counts must be used. Studies of seasonal variation in abundance of single species often require the conversion of the data to densities because detectability is likely to vary seasonally. Estimates of densities must also be used for comparisons of the relative abundance of different species, for example in community studies, because species differ in their detectability.

The literature on bird counting methods is extensive. There are reviews by Dawson (1985) and Verner (1985) and collections of many useful papers in the proceedings of both the Asilomar Symposium (Ralph and Scott 1981) and of meetings of the International Bird Census Committee (Oelke 1980; Purroy 1983; Taylor *et al.* 1985; Blondel and Frochot 1987). Information on the practical application of census techniques is given by Bibby *et al.* (1990).

7.2.5 Measures of breeding performance

The main information needed to assess the success of nesting attempts is the date of laying the first egg, the size of the completed clutch, and the survival of eggs and young between laying and fledging. The latter is usually broken down into separate estimates of survival for laying, incubation and chick rearing periods. A distinction is made between partial losses, which reflect hatching failure or starvation of chicks, and losses of whole nests, which usually result from predation or desertion. This information can easily be obtained by visiting nests and recording their contents. However, it is difficult to establish the number of nesting attempts made by individual pairs during the season without intensive studies of marked birds, which is rarely practical for routine monitoring.

The main difficulty in estimating nest survival is that most nests are not found until after the first egg is laid. The sample of nests under study is thus inevitably biased as all had survived up to the time when they were found. In such a sample the proportion of nests from which young fledged will give an overestimate of success. Much of this problem can be overcome by calculating average daily survival rates over the periods during which nests were observed (Mayfield 1961, 1975). This approach has been developed for whole nests within a formal statistical framework (Hensler and Nichols 1981; Hensler 1985; Bart and Robson 1982; see Pollock and Cornelius 1988 for an alternative approach). It can also be used to estimate the survival of individual eggs or young although further statistical development is needed here.

7.2.6 Measuring survival rates

Annual survival rates can be estimated from recovery records of ringed birds that are found dead. Survival rates are calculated from the ratios of recovered birds known to have been alive at the start of a particular time interval (because they were recovered subsequently) to those known to have been alive at the start of the previous time interval. Corrections must be made for annual variations in reporting rates (number reported/number dying), which can also be calculated from the recovery data if the numbers ringed in each year are known. Survival rates are calculated separately for first year and older birds because young birds usually survive less well than older ones. The analyses require separate sets of recoveries for birds ringed in their first year and as adults, because it is also likely that first year and older birds have different reporting rates. A range of statistical models for the analysis of such data has been developed, incorporating maximum-likelihood estimation of survival and reporting rates, goodness-of-fit tests to models and likelihood ratio tests between models (White 1983; Brownie *et al.* 1985; North 1987, 1990).

Survival rates can also be estimated by mark-recapture methods using live recaptures of ringed birds (Buckland and Baillie 1987; Peach *et al.* 1990). The method provides minimum estimates of survival because individuals that have emigrated permanently from the study area are considered to be dead. Adults of many species are faithful to their breeding sites and the method can provide useful measures of adult survival rates. Young birds are less likely to return to their natal areas to breed so juvenile survival rates may be severely underestimated, though juvenile return rates may sometimes provide a useful index of survival. Mark-recapture methods are particularly useful for species such as warblers, for which it is possible to obtain high numbers of recaptures but for which the chances of ringed birds being recovered as dead are extremely low. Sophisticated analytical methods for mark-recapture data have been developed along similar lines to those available for ringing recoveries (Clobert *et al.* 1987).

7.3 MONITORING SCHEMES IN THE UK AND IRELAND

7.3.1 The Common Birds Census and the Waterways Bird Survey

These schemes involve territory mapping and are designed to provide annual indices of population levels. The Common Birds Census (CBC) was started in 1962 and covers mainly farmland and woodland. The Waterways Bird Survey (WBS) started in 1974 and covers linear waterways (Carter 1989a). The censuses are carried out mainly by volunteers but territory clusters are determined by BTO staff analysts. About 100 farmland plots, 100 woodland plots and 120 waterways plots are covered each year.

Annual changes in population size are calculated as percentages from the summed territory counts for all plots that are covered in the same way in consecutive years. Confidence limits are calculated using the formulae given by Baillie *et al.* (1986). Population changes are published annually for 60 species censused by the CBC and for 18 species censused by the WBS. Annual changes greater than 30% have an 80% probability (power = 0.80) of being detected at a significance level of 5% ($P < 0.05$) for about 41 of the species censused by the CBC. For many of these species considerably smaller changes have a high probability of detection. The population index is then calculated by applying successive annual changes (the chain method) to a value of 100, set for a datum year (previously 1966, now 1980, Figure 7.2). Thus,

$$I_{t+1} = I_t \, \Sigma \, P_{i(t+1)}/\Sigma P_{it}$$

where I_t = population index for year t
  P_{it} = number of territory clusters on the *i*th plot in
    year t. Note that only plots covered in the same
    way in both years are included.

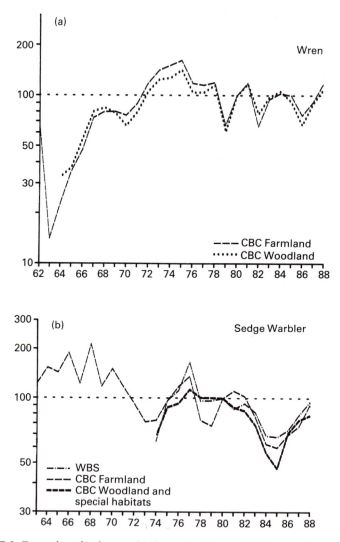

Figure 7.2 Examples of index results from the CBC and WBS (from Marchant *et al.* 1990).

(a) Farmland and woodland indices for Wren both show very similar patterns. Numbers decline sharply following severe winters, but show a rapid density-dependent recovery.

(b) WBS and CBC indices for Sedge Warbler. Because this species is only found on a limited number of plots an index for all plots other than farmland (i.e. woodland plus special habitats) has been calculated to make best use of the available data. All indices show the decline in this species, which is probably due to deterioration of the wintering areas in the drought-stricken Sahel region of Africa.

One potential problem with such indices is that random errors in the estimates of population changes can cause the index to drift away from its true value (Geissler and Noon 1981). Simulation studies suggest that this is unlikely to be a serious problem with the CBC data (Moss 1985), though it can make some interpretations difficult (Greenwood 1989).

Mountford (1982, 1985) has developed an alternative method for calculating index values which uses direct comparisons between all pairs of years. This gives greater precision from the same data but further development is required before it can be applied to long runs of CBC data.

The methods used by the CBC have been validated extensively (O'Connor and Fuller 1984). Early studies showed consistency of annual changes on different plots (Taylor 1985; Bailey 1967). A major study in which the same plot was censused by four observers and the maps analysed by three staff analysts (the Aston Rowant experiment; O'Connor 1981), showed that BTO analysts were consistent in their assessments of numbers of territories. Observers differed in their census efficiency; more experienced observers spent more time in the field and recorded more registrations (Figure 7.3). Despite this, percentage changes between years did not differ between observers, because counts in both years were affected equally by observer efficiency. Similar studies have been carried out for the WBS (Taylor 1983, 1985).

Participants in the CBC and WBS select their own study plots, although BTO staff ensure that new plots are suitable for inclusion in the scheme. Plots

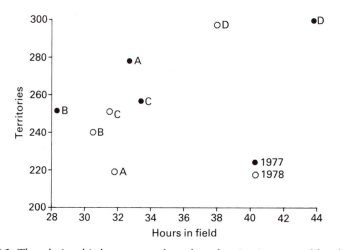

Figure 7.3 The relationship between total number of territories assessed for all species and time spent in the field for a constant ten census visits. Letters represent different observers. From O'Connor (1981).

should ideally be selected at random from the habitats of interest to ensure that they are representative but such a system would be unlikely to attract many volunteer observers. As a result, the geographical coverage of all BTO site-based monitoring reflects the geographical distribution of observers (Figure 7.4). Southern and eastern Britain (east of National Grid line 3000 and south of line 5000) is well covered by CBC plots and within this area the land-use categories of farmland CBC plots were similar to the statistics for the region as a whole (Fuller *et al.* 1985). The cropping patterns recorded on English farmland CBC plots also reflect closely the variation in the agricultural statistics compiled by MAFF. Studies of this kind have not yet been carried out for woodland CBC or WBS plots.

The long-term population trends shown by the CBC and WBS are reviewed by Marchant *et al.* (1990) and annual estimates of population changes are published in *BTO News* (e.g. Carter 1989b; Marchant and Whittington 1989).

7.3.2 The Nest Record Scheme

The Nest Record Scheme uses a network of around 1000 volunteers to gather data on the breeding performance of birds. Participants use standard cards to record the details of the nests that they find. The cards are designed to promote standardisation of recording and efficient computerisation. Participants are encouraged to visit nests every few days to record the progress of the nesting attempt. County, grid-reference, altitude, habitat and nest site are recorded in addition to the nest contents at each visit, the success of the nest and the reasons for failure if known. Over 800 000 nest histories have been recorded since the scheme was started in 1939, 200 000 of which are computerised. Standard computer programmes are used to calculate laying date, clutch size, brood size, egg survival and chick survival from the data on the cards.

Since the 1988 breeding season, nest records data for a range of species have been used to monitor annual changes in breeding performance (e.g. Crick 1989). Over 30 000 cards are submitted each year, and this level of recording will eventually allow annual monitoring of the breeding

---

Figure 7.4 The distribution of study plots used by BTO site-based monitoring schemes.
(a) Common Birds Census plots used in 1988
   ● Farms ■ Woods
(b) Waterways bird survey plots used in 1986
(c) Constant Effort Sites used in 1988
   ▲ Wetlands ● Dry scrub ■ Woods and other habitats

performance of about 68 species. Sampling within this scheme is inevitably non-random with respect to geographical location and habitat, although this can be overcome to a limited extent by encouraging observers to submit records from particular regions or habitats. Analyses of annual changes must take account of any variations in breeding performance between habitats or geographical regions. They must also take account of the decline in observer effort and nest finding efficiency through the season, which often result in late nests being under-represented in the sample (Morgan and Davis 1977).

7.3.3 Constant Effort Sites

The Constant Effort Sites Scheme (CES) uses standardised summer mist-netting at wetland, scrub and woodland sites to provide indices of abundance of adults and juveniles and estimates of adult survival. The net sites used and the duration of the netting sessions are the same in each year. Participants visit their sites once in each of 12 ten-day periods between the beginning of May and the end of August. All birds captured are ringed and classified as either adults or juveniles.

Paired comparisons using all sites covered in pairs of consecutive years are used to calculate percentage changes and index values as in the CBC. Comparisons are based on the number of individuals caught during the season, irrespective of how many times an individual was handled, and are calculated for adults and juveniles separately.

Survival rates are monitored using data from between-year recaptures of adults. Constant effort netting simplifies the analysis of survival rates by allowing the use of models where the capture probability does not vary between years (Peach *et al.* 1990). The use of these constant capture probabilities allows more precise estimates of survival rates than could be obtained from a variable netting regime with an equivalent amount of catching effort. The CES scheme started as a pilot project in 1981 and has been run as a full BTO scheme since 1986. Over 90 sites were included in the scheme by 1989 (Figure 7.4). Estimates of changes in adult numbers and productivity of 23 species are published annually in *BTO News* (Peach and Baillie 1989).

7.3.4 The BTO Ringing Scheme

The Ringing Scheme gathers data which can be used to monitor the survival rates of birds as well as providing much of our information on bird migrations and movements. It also has the potential to provide extensive monitoring data on productivity, from ratios of juveniles to adults captured, and on dispersal.

The Ringing Scheme has been in operation since 1909 and by 1987 over 19 million birds had been ringed and over 400 000 recoveries reported (Mead and Clark 1988). During the 1980s about 800 000 birds were ringed each year and about 14 000 recoveries reported. These data should make it possible to monitor the annual survival rates of at least 28 species. All recovery data are held on computer file and recoveries are routinely computerised. The ringing data are not currently computerised unless a bird is recovered, which severely limits the analyses that can be performed (see Section 7.2.6). Full computerisation of the ringing data is planned for the 1990s. Since 1985 information on the summer ringing of 22 species, categorised by age of bird and region of ringing, has been summarised by ringers and recorded on special forms (age-specific totals lists, Baillie and Green 1987). This will allow monitoring of annual survival rates to be started in the early 1990s. As with the Nest Record Scheme, it is impossible to control sampling strategies. Analyses must therefore take account of geographical and habitat-related variations in survival and recovery rates.

7.4 INTEGRATED POPULATION MONITORING

Until recently, 'monitoring' of populations of common birds in the UK and Ireland has been at the level of surveillance, with changes in the CBC index being reported annually. An Integrated Population Monitoring Programme is now being developed by the BTO to provide a true monitoring capability for bird populations (Baillie 1990). Its objectives are as follows.

1. To establish thresholds that will be used to notify conservation bodies of requirements for further research or conservation action.
2. To identify the stage of the life cycle at which changes are taking place.
3. To provide data that will assist in identifying the causes of population changes.
4. To distinguish anthropogenic changes in populations from 'natural' population fluctuations.

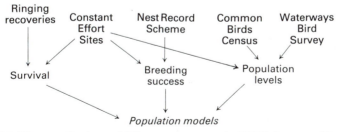

Figure 7.5 The contributions of different schemes to the BTO's Integrated Population Monitoring Programme.

The programme will combine data from the BTO schemes outlined above to provide annual indices of population size, productivity and survival rates, and to develop population models (Figure 7.5). Patterns of variation in the long-term data, including relationships with readily measured environmental variables such as winter temperatures (Greenwood and Baillie in press), will be used to establish action thresholds (objective 1, above). For migratory species it may be important to consider environmental conditions in their passage and wintering areas, such as the droughts in the Sahel (Winstanley *et al.* 1974). These data will ultimately make possible the development of predictive population models which incorporate both environmental and density-dependent effects. The models will be used to compare observed population trends with changes expected on the basis of past data. For example, the breeding populations of some small passerines decline following severe winters (Figure 7.6) and so population declines at such times need not give cause for concern. However, if populations decline following mild winters, or if they fail to show the normal density-dependent recovery following declines caused by cold weather, then explanations should be sought. Models are simplifications of reality and it will always be important to screen trends in individual population variables in addition to reviewing model predictions.

7.5 DISCUSSION

7.5.1 The future development of monitoring in the UK and Ireland

The systems for bird monitoring used by the BTO are undergoing rapid development. There is scope for improvements in the gathering of field data, in the computer systems used to store and analyse the data, and in methods of analysis and modelling.

The most severe gap in the field data is the restricted geographical, habitat

---

Figure 7.6 Examples of the results of integrated studies that will contribute to the development of population models as part of the Integrated Population Monitoring Programme.

(a) Relationship between the percentage change in the farmland CBC index for Wren and the number of freezing days in the preceeding Winter, averaged over 20 weather stations throughout Britain. Data from 1962–87, $r = -0.68$, $P < 0.001$, percentage change arcsin transformed. An example of an environmental variable which accounts for fluctuations in the population.

(b) The relationship between clutch size (from the BTO Nest Record Scheme) and woodland breeding population level (from the woodland CBC) for Wrens in Britain. Data for 1964–80, $r = -0.72$, $P < 0.002$. From O'Connor and Fuller (1985). An example of a density-dependent response which will contribute to regulation of the population.

and species coverage of current census schemes. The CBC and WBS provide good population indices, based on territory mapping, for around 70 species but they cannot be expanded greatly because of both the difficulties in recruiting fieldworkers in poorly covered regions and the considerable staff time required for map analysis. Methods other than territory mapping are

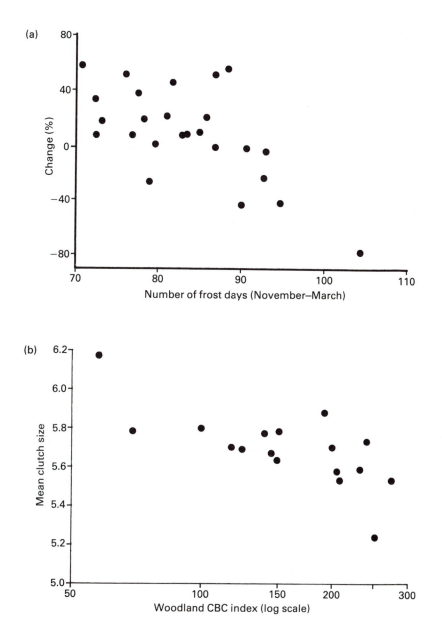

therefore being evaluated. Point counts and transects provide less precise information on the bird populations of individual study plots, but they require less time commitment from individual observers and facilitate efficient processing of the data by computer.

Periodic reviews are essential for ensuring that long-term monitoring schemes continue to meet their objectives and to use the best available methods. Reviews of BTO schemes are normally carried out by a group of independent scientists and staff. Reviews of the CBC (1983), CES (1985), Nest Record Scheme (1987) and Ringing Scheme (1987) objectives have been carried out and a group considering Integrated Population Monitoring will operate during the early 1990s.

Perhaps the most important development will be to carry out further integrated analyses of population dynamics (O'Connor 1980a and b; Baillie 1990), to provide a basis for the development of the models needed for the Integrated Population Monitoring Programme: realistic interpretation of monitoring data depends on an understanding of the factors that give rise to normal population fluctuations.

7.5.2 Monitoring bird populations on a European scale

Birds are highly mobile species and do not respect national boundaries. British and Irish bird populations have much in common with those on the European mainland. They are affected by common weather systems and often share the same passage and wintering areas. They are also subject to many similar anthropogenic influences, for example, the proliferation in use of agrochemicals. At least eight European countries operate breeding bird censuses: Britain, the Netherlands and Sweden use territory mapping while Sweden, Denmark, Czechslovakia, the Estonian CSSR, Finland and France use point counts (Hustings 1988). Britain and France operate constant effort mist-netting schemes during the breeding season and a late summer constant-effort netting programme is operated on a few continuously-manned sites in West Germany (Berthold *et al.* 1986). Co-ordination of bird census work in Europe and the exchange of results is carried out through the International Bird Census Committee (IBCC). There is also good co-ordination of ringing information through the European Union for bird Ringing (EURING). There are cogent ecological reasons for monitoring bird populations on a European scale, and substantial advances in this area can be expected during the 1990s.

ACKNOWLEDGEMENTS

My post at the British Trust for Ornithology is financed by the Nature Conservancy Council. I thank the staff of the BTO Populations Research

Department for assistance with many aspects of this work, and Drs Barrie Goldsmith, Jeremy Greenwood, Will Peach and Helen Smith for constructive criticism of earlier drafts.

REFERENCES

Anon (1969) Recommendations for an international standard for a mapping method in bird census work, *Bird Study*, **16**, 249–54.

Bailey, R.S. (1967) An index of bird population changes on farmland, *Bird Study*, **14**, 195–209.

Baillie, S.R. (1990) Integrated population monitoring of breeding birds in Britain and Ireland, *Ibis*, **132**, in press.

Baillie, S.R. and Green, R.E. (1987) The importance of variation in recovery rates when estimating survival rates from ringing recoveries, in *Ringing Recovery Analytical Methods, Acta Ornithologica*, **23**, 41–60.

Baillie, S.R., Green, R.E., Boddy, M. and Buckland, S.T. (1986) *An evaluation of the Constant Effort Sites Scheme*, BTO Research Report 21, British Trust for Ornithology, Tring.

Bart, J. and Robson, D.S. (1982) Estimating survivorship when the subjects are visited periodically, *Ecology*, **63**, 1078–90.

Berthold, P., Fliege, G., Querner, U. and Winkler, H. (1986) The development of songbird populations in central Europe: Analysis of trapping data, *Journal für Ornithologie*, **127**, 397–437.

Best, L.B. (1975) Interpretational errors in the 'mapping method' as a census technique, *Auk*, **92**, 452–60.

Bibby, C.J., Burgess, N. and Hill, D.A. (1990) *A guide to census techniques*, Academic Press, London, in press.

Blondel, J. and Frochot, B. (eds) (1987) *Bird census and atlas studies, Acta Oecologica (Oecologia Generalis)*, **8**, No. 2.

Brownie, C., Anderson, D.R., Burnham, K.P. and Robson, D.S. (1985) *Statistical inference from band recovery data – a handbook*, US Fish and Wildlife Service Resource Publication No. 156, Washington.

Buckland, S.T. (1987) On the variable circular plot method of estimating animal density, *Biometrics*, **43**, 363–84.

Buckland, S.T. and Baillie, S.R. (1987) Estimating bird survival rates from organised mist-netting programmes, in *Ringing Recovery Analytical Methods, Acta Ornithologica*, **23**, 89–100.

Burnham, K.P., Anderson, D.R. and Laake, J.L. (1980) Estimation of density from line transect sampling of biological populations. *Wildlife Monographs*, **72**.

Carter, S.P. (1989a) The Waterways Bird Survey of the British Trust for Ornithology: an overview, *Regulated Rivers: Research and Management*, **4**, 191–7.

Carter, S.P. (1989b) Waterways Bird Survey 1987–8 population changes, *BTO News*, **161**, 10–11.

Clobert, J., Lebreton, J.D. and Allaine, D. (1987) A general approach to survival rate estimation by recaptures or resightings of marked birds, *Ardea*, **75**, 133–42.

Crick, H.Q.P. (1989) Breeding birds in 1988. *BTO News*, **164**, 9–12.

Dawson, D.G. (1981a) Experimental design when counting birds, in *Estimating numbers of terrestrial birds* (eds C.J. Ralph and J.M. Scott), *Studies in Avian Biology*, **6**, 392–8.

Dawson, D.G. (1981b) The usefulness of absolute ('census') and relative ('sampling' or 'index') measures of abundance, in *Estimating numbers of terrestrial birds* (eds C.J. Ralph and J.M. Scott), *Studies in Avian Biology*, **6**, 554–8.

Dawson, D.G. (1985) A review of methods for estimating bird numbers, in *Bird Census and Atlas Studies* (eds K. Taylor, R. J. Fuller and P.C. Lack), British Trust for Ornithology, Tring, pp. 27–33.

Emlen, J.T. (1971) Population densities of birds derived from transect counts, *Auk*, **88**, 323–42.

Emlen, J.T. (1977) Estimating breeding season bird densities from transect counts, *Auk*, **94**, 455–68.

Enemar, A., Sjostrand, B., and Svensson, S. (1978) The effect of observer variability on bird census results obtained by a territory mapping technique, *Ornis Scandinavica*, **9**, 31–9.

Falls, J.B. (1981) Mapping territories with playback: an accurate census method for songbirds, in *Estimating numbers of terrestrial birds* (eds C.J. Ralph and J.M. Scott), *Studies in Avian Biology*, **6**, 86–91.

Fuller, R.J. and Langslow, D.R. (1984) Estimating the numbers of birds by point counts: how long should the counts last?, *Bird Study*, **31**, 195–202.

Fuller, R.J., Marchant, J.H. and Morgan, R.A. (1985) How representative of agricultural practice in Britain are Common Birds Census farmland plots?, *Bird Study*, **32**, 56–70.

Geissler, P.H. and Noon, B.R. (1981) Estimates of avian population trends from the North American Breeding Bird Survey, in *Estimating numbers of terrestrial birds* (eds C.J. Ralph and J.M. Scott), *Studies in Avian Biology*, **6**, 42–51.

Greenwood, J.J.D. and Baillie, S.R. (1990) Effects of density dependence and weather on population changes of English passerines; using the non-experimental paradigm, *Ibis*, in press.

Hensler, G.L. (1985) Estimation and comparison of functions of daily nest survival probabilities using the Mayfield method, in *Statistics in Ornithology* (eds B.J.T. Morgan and P.M. North), Springer-Verlag, Berlin, pp. 289–97.

Hensler, G.L. and Nichols, J.D. (1981) The Mayfield method of estimating nesting success: a model, estimators and simulation results, in *Wilson Bulletin*, **93**, 42–53.

Hustings, F. (1988) *European monitoring studies on breeding birds*, Samenwerkende Organisaties Vogelonderzoek Nederland, Beek.

Jarvinen, O. and Vaisanen, R.A. (1975) Estimating relative densities of breeding birds by the line transect method, *Oikos*, **26**, 316–22.

Jarvinen, O. and Vaisanen, R.A. (1976) Finnish line transect censuses, *Ornis Fennica*, **53**, 115–18.

Marchant, J.H. (1981) Residual edge effects with the mapping bird census method, in *Estimating numbers of terrestrial birds* (eds C.J. Ralph and J.M. Scott), *Studies in Avian Biology*, **6**, 488–91.

Marchant, J.H., Hudson, R.W., Carter, S.P. and Whittington, P.A. (1990) *Population trends in British breeding birds*, British Trust for Ornithology, Tring.

Marchant, J.H. and Whittington, P.A. (1989) 1987–88 CBC index report, *BTO News*, **162**, 9–12.

Mayfield, H.F. (1961) Nesting success calculated from exposure, *Wilson Bulletin*, **73**, 255–61.

Mayfield, H.F. (1975) Suggestions for calculating nest success, *Wilson Bulletin*, **87**, 456–66.

Mead, C.J. and Clark, J.A. (1988) Report on bird ringing in Britain and Ireland for 1987, *Ringing and Migration*, **9**, 169–204.

Merikallio, E. (1958) Finnish birds, their distribution and numbers, *Fauna Fennica*, **5**, 1–181.

Morgan, R.A. and Davis, P.G. (1977) The number of broods reared by Stonechats in Surrey, *Bird Study*, **24**, 229–32.

Moss, D. (1985) Some statistical checks on the BTO Common Birds Census Index – 20 years on, in *Bird Census and Atlas Studies* (eds K. Taylor, R.J. Fuller and P.C. Lack), British Trust for Ornithology, Tring, pp. 175–9.

Mountford, M.D. (1982) Estimation of population fluctuations with application to the Common Bird Census, *Applied Statistics*, **31**, 135–43.

Mountford, M.D. (1985) An index of population change with application to the Common Bird Census, in *Statistics in Ornithology* (eds B.J.T. Morgan and P.M. North), Springer-Verlag, Berlin, pp. 121–32.

Newton, I. (1979) *Population ecology of raptors*, T. and A.D. Poyser, Berkhamsted.

Newton, I. (1986) *The Sparrowhawk*, T. and A.D. Poyser, Calton.

North, P.M. (1977) A novel clustering method for estimating numbers of bird territories, *Applied Statistics*, **26**, 148–55.

North, P.M. (1979) A novel clustering method for estimating numbers of bird territories: an addendum, *Applied Statistics*, **28**, 300–1.

North, P. M. (ed.) (1987) *Ringing Recovery Analytical Methods*. Proceedings of the Euring Technical Conference and meeting of the Mathematical Ecology Group of the Biometric Society (British Region) and British Ecological Society. *Acta Ornithologica*, **23**, 1–175.

North, P.M. (1990) *Proceedings of the second Euring Technical Conference, The Ring*, in press.

O'Connor, R.J. (1980a) Pattern and process in Great Tit *Parus major* populations in Britain, *Ardea*, **68**, 165–83.

O'Connor, R.J. (1980b) Population regulation in the Yellowhammer *Emberiza citrinella* in Britain, in *Bird Census Work and Nature Conservation* (ed. H. Oelke), Dachverband Deutscher Avifaunisten, Gottingen, pp. 190–200.

O'Connor, R.J. (1981) The influence of observer and analyst efficiency in mapping method censuses, in *Estimating numbers of terrestrial birds* (eds C.J. Ralph and J.M. Scott), *Studies in Avian Biology*, **6**, 372–6.

O'Connor, R.J. and Fuller, R.J. (eds) (1984) *A re-evaluation of the aims and methods of the Common Birds Census*, B.T.O. Research Report 15, British Trust for Ornithology, Tring.

O'Connor, R.J. and Fuller, R.J. (1985) Bird population responses to habitat, in *Bird Census and Atlas Studies* (eds K. Taylor, R.J. Fuller and P.C. Lack), British Trust for Ornithology, Tring, pp. 197–211.

Oelke, H. (ed.) (1980) *Bird census work and nature conservation*, Dachverband Deutscher Avifaunisten, Gottingen.

Peach, W.J. and Baillie, S.R. (1989) Population changes on Constant Effort Sites 1987–1988, *BTO News*, **161**, 12–13.

Peach, W.J., Buckland, S.T. and Baillie, S.R. (1990) Estimating survival rates using mark-recapture data from multiple ringing sites. Proceedings of the second EURING Technical Conference, *The Ring*, in press.

Pollock, K.H. and Cornelius, W.L. (1988) A distribution-free nest survival model, *Biometrics*, **44**, 397–404.

Purroy, F.J. (ed.) (1983) *Bird census and Mediterranean landscape*, University of Leon, Leon.

Ralph, C.J. and Scott, J.M. (eds) (1981) *Estimating numbers of terrestrial birds. Studies in avian biology*, **6**, Cooper Ornithological Society, Las Cruces.

Ratcliffe, D.A. (1980) *The Peregrine Falcon*, T. and A.D. Poyser, Calton.

Reynolds, R.T., Scott, J.M. and Nussbaum, R.A. (1980) A variable circular-plot method for estimating bird numbers, *Condor*, **82**, 309–13.

Svensson, S.E. (1979) Census efficiency and number of visits to a study plot when estimating bird densities by the territory mapping method, *Journal of Applied Ecology*, **16**, 61–8.

Taylor, K. (1983) A comparison of inter-observer and inter-analyst agreement in the BTO Waterways Bird Survey, *Bird Census and Mediterranean Landscape* (ed. F.J. Purroy), University of Leon, Leon, pp. 18–22.

Taylor, K. (1985) A further validation of the Waterways Bird Survey technique, *Bird Census and Atlas Studies* (eds K. Taylor, R.J. Fuller and P.C. Lack), British Trust for Ornithology, Tring, pp. 181–90.

Taylor, K., Fuller, R.J. and Lack, P.C. (eds) (1985) *Bird Census and Atlas Studies*, British Trust for Ornithology, Tring.

Taylor, S.M. (1965) The Common Birds Census – some statistical aspects, *Bird Study*, **12**, 268–86.

Tomialojc, L. (1987) On the aims and strategy of the International Bird Census Committee, *Bird census and atlas studies* (eds J. Blondel and B. Frochot), *Acta Oecologica (Oecologia Generalis)*, **8**, 93–102.

Verner, J. (1985). Assessment of counting techniques, in *Current Ornithology, Volume 2* (ed. R.F. Johnston), Plenum Press, New York, pp. 247–302.

White, G.C. (1983). Numerical estimation of survival rates from band-recovery and biotelemetry data, *Journal of Wildlife Management*, **47**, 716–28.

Wiens, J.A. (1969) An approach to the study of ecological relationships among grassland birds, *Ornithological Monographs*, **8**.

Winstanley, D., Spencer, R. and Williamson, K. (1974) Where have all the whitethroats gone?, *Bird Study*, **21**, 1–14.

— 8

National species distribution surveys

PAUL T. HARDING

8.1 INTRODUCTION

8.1.1 Background

A chapter on *surveys* may seem anomalous in a volume on *monitoring* where the differences between the two activities are so clearly defined. However, national species distribution surveys provide one of the fundamental baselines of information against which monitored information may be put in a wider context, in particular, information on the occurrence of species.

In this chapter various types of national species distribution surveys and the results of these surveys will be described especially those involving the Biological Records Centre (BRC).

The British Isles has traditionally been regarded as a biogeographical unit and, for the past 25 years, BRC has been seen as the focal point for species distribution surveys covering Britain and Ireland. The formation of the Biological Resources Data Bank for Ireland within the Wildlife Service (Office of Public Works) in the Republic of Ireland, and the planned database for Northern Ireland, may call for a review of coverage of Ireland as a whole by BRC. Throughout this chapter the work of 'national' surveys should be understood to cover the United Kingdom of Great Britain and Northern Ireland, the Isle of Man, the Channel Islands and the Republic of Ireland.

In the British Isles we are fortunate in having a tradition of amateur natural history. This tradition has been successfully harnessed, fostered and developed by motivating many thousands of volunteers with specialist knowledge to contribute to species distribution surveys.

Few other areas have a tradition as strong as ours, but many countries around the world have set up data centres similar to BRC, each of which is dealing with the collection and use of data in slightly different ways.

8.1.2 The aims of surveys

National distribution surveys aim to collect data to answer four basic questions:

- What species?
- Where was it found?
- When was it found?
- Who was responsible for the record?

Within each of these fields there are many subdivisions of detailed information which may be recorded, such as precise grid references, locality names, habitats and numbers of individuals. Harding in Copp and Harding (1985) attempted to describe the range of topics that could be covered by a record. A more general definition of a 'biological recording' can be given as follows.

> Biological recording is defined as the collection, collation, storage, dissemination and interpretation of spatially and temporally referenced information on the occurrence of biological taxa, assemblages and biotopes. Basic information on occurrence is normally augmented and amplified with a variety of related biological, environmental and administrative information. Biological recording normally excludes information on agricultural, horticultural or forestry crops, and agricultural, domestic or captive stock, except where it may concern wildlife, biotopes or the management of semi-natural areas.

The overriding aim of most volunteers concerned with national species distribution surveys is mapping the distribution of species and the pursuit of taxonomic and ecological knowledge about their specialist taxonomic group. The conservation of species and protection of sites is also an important motivation for many specialists.

8.2 TYPES OF SURVEYS

8.2.1 History of surveys in the British Isles

Attempts to catalogue the flora and fauna of individual countries, or larger areas, have a long history, with claims being made for several Greek and Roman 'philosophers' as the father of scientific biology. In Britain, we look to John Ray (1627–1705) as the father of the scientific investigation of our flora and fauna. In his studies, Ray was distinctly aware of the need to document where species occurred (both habitat and location) as well as what they looked like. For example, in *Synopsis methodica Stirpium Britannicum* 1724 (Stearn 1973) Ray drew upon the local knowledge of others to gather available national information on the occurrence of species.

Species distributions in the British Isles have been summarised using subdivisions of administrative counties, known as 'vice-counties', devised by Watson (1859, 1873–4) for Britain and Praeger (1901) for Ireland. Both authors dealt with vascular plants, but their systems were soon adopted by authors for other groups. Publication of summaries of the national distribution of species, using vice-county maps, 'typomaps' (see Praeger 1906) or tables, proliferated.

The display of information on the distribution of species entered a new phase in 1951. In that year the Botanical Society of the British Isles (BSBI) decided to use the expertise of its membership to collect information on the distribution of vascular plants in the British Isles and to map the resultant data using the 10 km squares of the British National Grid as the basic recording and mapping unit (Clapham 1951). The resultant *Atlas of the British Flora* (Perring and Walters 1962) provided a model on which most subsequent species distribution surveys, both in the British Isles and abroad, have been based.

The essential elements of the BSBI model are as follows.

1. Volunteer specialists are used to collect the original data.
2. A small professional unit deals with data collation, data processing and map production.
3. Uniform cartographic cells are used to display species distributions on maps.

8.2.2 Biological Records Centre (BRC)

The national Biological Records Centre (BRC) at Monks Wood Experimental Station, near Huntingdon, was founded in 1964 on the methods, staff and data of the BSBI Atlas project. The original function of BRC was to map the distribution of species in the British Isles (Perring 1971) and to collect more detailed information on rare species.

The role of BRC developed to serve the then Nature Conservancy as a data-gathering and data-processing unit, but in 1973, the Nature Conservancy was split into the Nature Conservancy Council and the Institute of Terrestrial Ecology. Within the latter, BRC's role was modified and became concerned with links with the growing number of local records centres (see Chapter 9) and with fostering species recording and mapping in other European countries (Heath and Perring 1978).

Although the physical work of BRC changed only slightly, mainly in response to improved data-processing facilities, by the 1980s the *raison d'être* had undergone a philosophical change. Harding (1985) described the objectives of BRC thus:

1. to set up and operate a computerised data bank of information on the occurrence of plants and animals in the UK;
2. to maintain an archive of the original records from which the data bank was compiled;

3. to make these data available in a variety of forms for research, monitoring, nature conservation, education and general information.

In 1989, BRC was operating 64 national recording schemes covering some 16 000 species, with 4.5 million records on computer file. Distribution maps of over 6000 species have either been published (Harding 1989) or are in preparation.

In 1989, BRC became one of the component units of the Institute of Terrestrial Ecology's Environmental Information Centre. This new centre brought together expertise and technology in the use of ecological databases, remotely sensed data on land use and vegetation, digital mapping and geographical information systems.

The scale of the BRC database is governed by the resources it has available to it. These resources have fluctuated as successive governments have influenced research policies, priorities and funding. The importance of a national overview of the occurrence of species, such as can be provided by BRC, is becoming increasingly apparent as concern for wildlife and 'the environment' becomes a public and political issue.

8.2.3 Other surveys in the British Isles

Several major distribution surveys of the flora and fauna of the British Isles are conducted independently of BRC, although most had some contact with the Centre, for example for data processing, map production or publication.

The British Lichen Society began a mapping scheme in 1963 which has continued virtually unaltered for over 25 years. Some results have been published, in the *Lichenologist*, in papers and books, Volume 1 of an atlas (Seaward and Hitch 1982) and provisional atlases (Seaward 1984, 1985).

The British Trust for Ornithology has conducted 3 major national mapping schemes; of breeding birds (Sharrock 1976), of wintering birds (Lack 1986) and a resurvey of breeding birds (being made between 1988 and 1990).

The Wildfowl and Wetland Trust surveys the distribution of wildfowl throughout the British Isles, combining censuses with the geographical distribution of wintering flocks.

More information about monitoring birds is given in Chapter 7.

8.2.4 European and international surveys

International collaboration to map the distribution of species on, for example, a pan-European scale has been afflicted by lack of resources. Despite this lack, several bold initiatives have been taken and mapping projects for vascular plants, mammals, herptiles, breeding birds, moths and myriapods are currently operating. Other projects, including butterflies, have been proposed.

Most of the international projects are at a basic level, to map summarised data using large grid cells (for example 50 km squares of the Universal Transverse Mercator grid). Data are collated in summarised form from national centres (where they exist or hold relevant data) and from national experts and co-ordinators. Broad-brush surveys based on secondary sources have limited value, but they permit an accurate, but often incomplete, assessment of the geographical range of species.

The existence of up-to-date information sources in European countries is essential for international organisations, such as the Council of Europe and the European Community, which are concerned with wildlife legislation (e.g. the Bern Convention, the Ramsar Agreement and the EC Habitats Directive). There is some evidence that the existence of international species distribution projects helps to emphasise the importance of wildlife and the value of national data centres and national surveys. In 1987, the Committee of Ministers of the Council of Europe recommended (R(87)13) that local and national environmental data centres should be set up and maintained in member states.

8.3 SURVEY METHODS

The national (i.e. British Isles) species distribution surveys co-ordinated by BRC are based on *recording schemes* for individual taxonomic groups (e.g. vascular plants, butterflies or mammals). The exception to this rule is the scheme for plant galls which covers all gall-forming taxonomic groups and their host plants.

8.3.1 Setting up a recording scheme

Four factors are seen as important for a scheme to be successful.

1. A volunteer national scheme organiser and/or a network of volunteer regional organisers.
2. Volunteer specialists to record for the scheme.
3. Readily accessible identification guides.
4. A practical selection of species to be covered (neither too many nor too few, and avoiding mixes of species which require vastly different survey techniques).

The first two factors are essential, but a few schemes have been moderately successful without guides or with seemingly impractical mixes of species (e.g. aquatic and terrestrial). In several cases, national organisers have undertaken to identify all or most specimens associated with records.

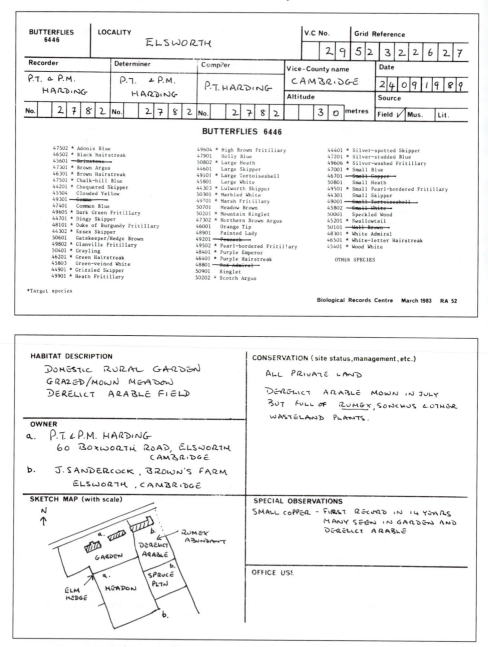

Figure 8.1 Species list/site card (RA 52-Butterflies). An example of the commonest type of card used for recording the occurrence of a taxonomic group at a locality (front, top; reverse, bottom).

8.3.2 Collection of data

All schemes run in collaboration with BRC use recording cards specially designed for the particular group (e.g. butterflies – Figure 8.1) and/or one or more general cards (multiple records of one species – Figure 8.2 or individual records for scarce species or unusual records – Figure 8.3). Cards are provided free to those involved with recording schemes and are distributed by BRC or the scheme organiser.

The volunteer recorders use the cards to submit their records (observations) to BRC in a standard format. Such a standard is essential for data processing and data manipulation. Data, after some coding, can be input via a keyboard, to the BRC database. Some use has been made of optical mark/read (OMR) forms, but they proved unreliable and were not physically robust. However, the Secretariat de la Faune et de la Flore, in Paris, has made successful use of OMR forms for species mapping.

The increasing availability of home computers brings ever closer the day when a majority of records will be submitted in a computerized form by recorders to national schemes and BRC. Data transfer standards for BRC are being developed to guide potential contributors.

The volunteer specialists collect and record their individual taxonomic group and submit records, usually annually, to the regional or national organiser for that group. Data are stored with the national or regional organiser until BRC has resources to process and computerise the data.

8.3.3 Validation of data

There are three forms of validation in processing data for the national database at BRC.

(a) Taxonomic
This is carried out by the regional and national scheme organisers, sometimes in collaboration with other taxonomic specialists. This validation may involve the checking of specimens, descriptions or photographs. Further taxonomic checking may be necessary after data-processing when apparently anomalous records are sorted from the data.

(b) Geographic
Performed mainly at BRC, partly using computer techniques to compare grid references and vice-counties to identify gross errors. The use of computerised gazetteers in checking geographical information is being developed at BRC. There is a recurrent problem with detailed site-related data when there is variation in the way place names are expressed, particularly because

ORDER	ACULEATE	GENUS & SPECIES		PODALONIA		SUB-SPECIES	

| 6 | 4 | 5 | 2 | HYMENOPTERA | 1 | 1 | 3 | 0 | 1 | AFFINIS (Kirby) | |

COMPILER P.T. HARDING   SOURCE   (Collection/Reference)
ARCHER, M.E. 1989. Entomologist's mon. Mag.,
125 p 232.

| | | 4 | 7 | Fld. | Mus. | Lit. |

Grid Reference	V.-C.	Collector/ Recorder	Determiner	Locality	Date
52/79-71-	26	M.E. ARCHER 03	M.E. ARCHER 03	RAMPART FIELD	7/7/1987
52/843 842	26	,,	,,	THETFORD WARREN LODGE	14/8/1984
ditto	26	,,	,,	ditto	6/7/1987
52/758 728	26	,,	,,	CAVENHAM HEATH NNR	15/7/1983
ditto	26	,,	,,	ditto	8/7/1987
52/854 866	28	,,	,,	TWO MILE BOTTOM COMMON	9/7/1987
62/351 456	25	,,	,,	LOWER HOLLESLEY COMMON	11/7/1987
62/20-42-	25	,,	,,	NACTON COMMON	12/7/1987
62/47-68-	25	,,	,,	DUNWICH HEATH	13/7/1987

Figure 8.2 Single species card (GEN 7). Used for extracting records from herbaria, collections or publications.

conservation organisations frequently use non-standard names for protected sites.

(c) Input and editing

Even skilled computer staff make mistakes in keyboarding data and editing computer files! Their errors are checked at BRC using both automatic and manual techniques.

8.3.4 Storage and dissemination of data

The BRC database uses the ORACLE database management system on VAX computers operated by the Natural Environment Research Council Computer Services. Much of the manipulative work on datasets, before

Figure 8.3 (opposite) An individual record card (GEN 8). Used for the occurrence of rare or threatened species or of particularly notable records.

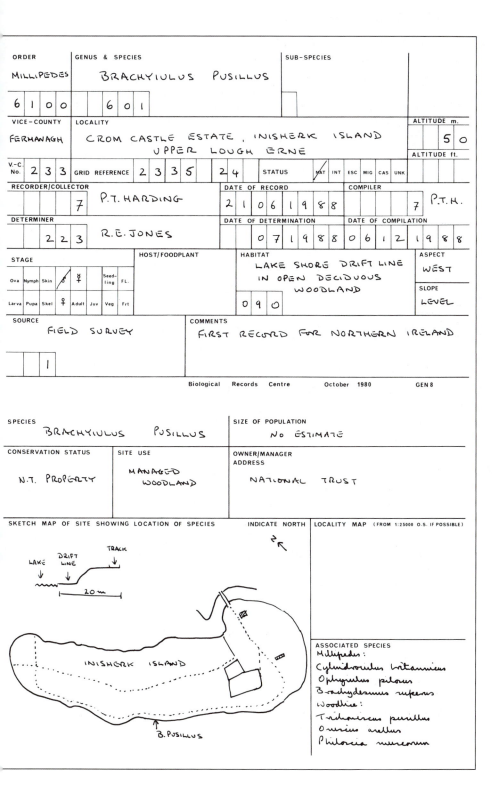

ORDER	GENUS & SPECIES	SUB-SPECIES	
MILLIPEDES	BRACHYIULUS PUSILLUS		
6 1 0 0	6 0 1		

VICE-COUNTY	LOCALITY	ALTITUDE m.
FERMANAGH	CROM CASTLE ESTATE, INISHERK ISLAND UPPER LOUGH ERNE	5 0

ALTITUDE ft.

| V.-C. No. | 2 3 3 | GRID REFERENCE | 2 3 3 5 | 2 4 | STATUS | NAT | INT | ESC | MIG | CAS | UNK | |

RECORDER/COLLECTOR	DATE OF RECORD	COMPILER
7 P.T. HARDING	2 1 0 6 1 9 8 8	7 P.T.H.

DETERMINER	DATE OF DETERMINATION	DATE OF COMPILATION
2 2 3 R.E. JONES	0 7 1 9 8 8	0 6 1 2 1 9 8 8

STAGE	HOST/FOODPLANT	HABITAT	ASPECT
Ova Nymph Skin ♂ ♀ Seedling FL. Larva Pupa Skel ♀ Adult Juv Veg Frt		LAKE SHORE DRIFT LINE IN OPEN DECIDUOUS WOODLAND 0 9 0	WEST SLOPE LEVEL

SOURCE	COMMENTS
FIELD SURVEY 1	FIRST RECORD FOR NORTHERN IRELAND

Biological Records Centre   October 1980   GEN 8

SPECIES	SIZE OF POPULATION
BRACHYIULUS PUSILLUS	NO ESTIMATE

CONSERVATION STATUS	SITE USE	OWNER/MANAGER ADDRESS
N.T. PROPERTY	MANAGED WOODLAND	NATIONAL TRUST

SKETCH MAP OF SITE SHOWING LOCATION OF SPECIES   INDICATE NORTH   LOCALITY MAP (FROM 1:25000 O.S. IF POSSIBLE)

LAKE   DRIFT LINE   TRACK
20 m

INISHERK ISLAND

B. PUSILLUS

ASSOCIATED SPECIES
Millipedes:
Cylindroiulus britannicus
Ophyiulus pilosus
Brachydesmus superus
Woodlice:
Trichoniscus pusillus
Oniscus asellus
Philoscia muscorum

incorporation in the ORACLE database, uses specially written software to deal with the requirement of individual datasets.

In common with any computerised database, information is available in computerised or hard-copy form (see Section 8.4.1). In addition, an archive of original record cards and other source material is maintained at BRC. This archive contains some additional information on species records which it would be impractical to incorporate in a national database.

Publication of results from national species distribution surveys is normally in the form of atlases or other publications including distribution maps. An up-to-date list of these publications was given by Harding (1989).

8.4 PRODUCTS OF SURVEYS

The products of surveys are both tangible (data, maps, publications, site and species protection) and intangible (increased scientific expertise, knowledge of the taxonomy and ecology of species, appreciation of the natural environment, personal fulfillment for individual volunteers). It would be impractical (and largely irrelevant in the context of this volume) to describe these intangible products – interesting and important as they may be.

8.4.1 Types of output

Output from the national database at BRC is available in a variety of forms and in two media – printed on paper, or in a computer-readable form, either as micro-computer disks or via the Joint Academic Network (JANET).

(a) Maps
Distribution maps can be prepared from the database for any one of the species for which data are held, or for any combination of species. Maps normally show the distribution of species in 10 km squares of the British and Irish National Grids (Figure 8.4). A variety of symbols may be used to differentiate between date periods of records or different types of records (e.g. validated or unconfirmed identification, native or introduced populations).

(b) Basic data
Available in a variety of forms such as all the records of a species, or a selection of species; all the records from a locality or a grid cell; all the records for some other data field (e.g. recorder, determiner, date, museum or literature source).

(c) Interpreted data
Data for a species or for the species at a site can be interpreted by BRC, or by relevant specialists to provide assessments of the local, national or

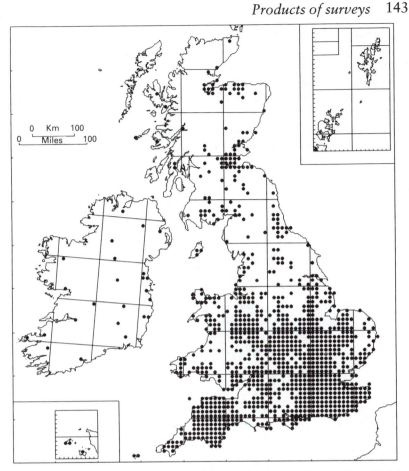

Figure 8.4 Species distribution map. *Epilobium ciliatum*, an alien species which has spread since it was first collected in 1891. ● All records.

international importance of the species recorded. Ecological and biological information and details of factors affecting the occurrence of species may be available. Within the ITE Environmental Information Centre, BRC has access to a geographical information system which will enable species information to be inter-related with other geographically-referenced environmental data such as climate, soils, geology and statutory site protection.

8.4.2 Uses of information

National species distribution surveys can provide a wide variety of information about the occurrence of species with several levels of complexity.

(a) Level 1   Location and date

This level of information allows the geographical range of species to be defined using distribution maps (Figure 8.4) or other semi-diagrammatic summaries (typomaps, tables etc.). The geographical range combined with temporal information, especially where the survey has been conducted over a long period of time, or has extracted earlier records from publications and collections, can be used to demonstrate changes in distribution over time (Figure 8.5). Geographical range, expressed as the number of 10 km squares

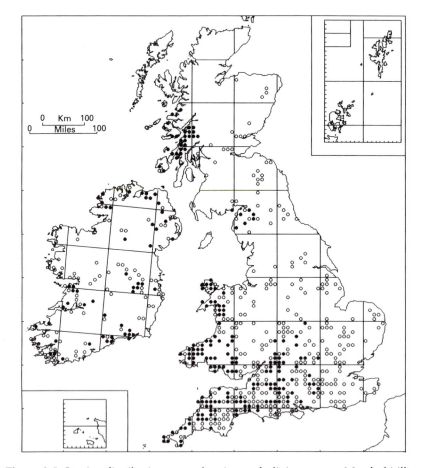

Figure 8.5 Species distribution map showing a declining range. Marsh fritillary butterfly (*Euphydryas aurinia*). ○ Records up to 1969, but not since. ● Records since 1969.

in which a species occurs, and an awareness of changes in distribution have been key components in assessing the status of rare and threatened species, for example in Red Data books (Perring and Farrell 1983; Shirt 1987).

The number of 10 km squares in which a species occurs, and the geographical range of species, are used as criteria in assessing the rarity of species for the selection of protected areas (SSSIs) (Nature Conservancy Council 1989). Using a simple scoring procedure, based on the number of threatened, rare or scarce species recorded at a site, the importance of a site is graded. Many factors, other than rarity, are also included in the criteria for selecting SSSIs.

(b) Level 2   Level 1 + Localities
Locality information, as either or both detailed grid references and accurate place names, provides a level of detail which is essential for those concerned with species and habitat protection, autecological research and taxonomic study. Such information is also essential for environmental assessment in planning.

(c) Level 3   Level 2 + 'Habitat'
Collecting 'habitat' information as part of distribution surveys should only be undertaken when there are clear uses for the information. For example, there is little point in collecting simplified habitat information (e.g. woodland, grassland, garden) for taxonomic groups for which habitats are already known (e.g. flowering plants or butterflies) or which are very mobile (e.g. adult dragonflies). Recording 'habitats' is most successful and productive where it is likely to advance knowledge of the habitat preferences of species and where this information is lacking or poorly understood, for example for woodlice (Harding and Sutton 1985).

(d) Level 4   Level 1 + Sexes, developmental stages or detailed dates
Understanding the biology of species, especially some invertebrates, can be advanced using data from national surveys. Recent analyses of data for dragonflies show differences in flight periods for some species, in relation to geographical range (Merritt, Moore and Eversham, in press). Several BRC schemes are collecting data on sexes and developmental stages to gather more information on the seasonal activity of sexes and to obtain proof of breeding of mobile species.

In many ways, the Breeding Birds (Sharrock 1976) and Wintering Birds (Lack 1986) surveys organised by the BTO are at this level. One of the successes of these surveys was in demonstrating differences and similarities in the ranges of birds during the breeding season and during the winter.

8.5 INTERPRETATION OF SPECIES DISTRIBUTIONS

Great care has to be taken in interpreting the results of national species distribution surveys, or any other attempt to assess the distribution of species. Problems may arise from three main causes which are described below, with examples. Despite these problems, meaningful interpretations of the occurrence of species can be made, for example to demonstrate changes in their distribution. Several of the examples of species cited below are described in greater detail by Preston and Eversham (in preparation).

8.5.1 Knowledge of taxonomy is unreliable or incomplete

(a) Taxonomic confusions

There are many examples where apparently good species, regarded as such for decades, are proved to be two or more species. For example, the moss *Anoectangium warburgii* was split (as a species new to science) from *Gymostomum calcareum* (Crundwell and Hill 1977), and the woodlouse *Haplophthalmus montivagus* was not recognised by British workers as being distinct from the commoner *H. mengei* (Hopkin and Roberts 1987).

(b) Closely related species

Although species have been recognised as distinct species, the reliable identification of specimens is only possible by a few experts. For example hawkweeds (*Hieracium*) and dandelions (*Taraxacum*) among the flowering plants and, among the moths, 5-spot burnets (*Zygaena* spp.) and some pugs (Eupithecia spp.).

(c) Overlooked species

Species whose presence, although known and recognised, is masked because assumptions have been made. For example, the scarce emerald damselfly *Lestes dryas* was regarded as extinct in Britain (but not in Ireland) between 1973 and 1982, but was 'refound' in 1983 and has since been recorded in over 18 10-km squares in Britain. In all probability it survived in several areas between 1973 and 1982, but was not sought at new sites because it had genuinely died out at some previously well known sites.

8.5.2 Bias in recording

(a)   Popular v neglected taxonomic groups

Some taxonomic groups are popular with 'amateur' specialists, (for example, birds, butterflies and flowering plants) whilst others have a very small following (for example, leeches, centipedes and myxomycetes). The popularity of a group affects the rate at which records accumulate, with recorders for less popular or neglected groups taking many decades to achieve even partial coverage of the British Isles (see also Section 8.5.2c below).

(b) Popular groups with obscure species

The ecology and behaviour of individual species may render them obscure and difficult to record using whatever methods are standard for that group. Hairstreak butterflies (subfamily *Theclinae*) are notoriously sedentary and unlikely to be encountered except with special techniques (for example observing the tree canopy in woodland, or recording eggs).

The club-tailed dragonfly *Gomphus vulgatissimus*, a riverine species, is not seen commonly in flight and was considered to be restricted to a few localities in England. Surveys for this species, based on detecting the cast exuviae (skins of nymphs) on emergent vegetation, have shown it to be much more widespread and numerous than originally thought.

(c) Distribution of recorders

One of the commonest criticisms levelled at national distribution surveys is that they are dependent on where recorders chose to record. Such criticism is justified in the earlier stages of a survey and with those surveys of groups which are 'neglected'. Recording is often an almost obsessive enthusiasm for many specialists and, as a result of their enthusiasm, records are obtained from far and wide. Coverage will always be best where specialists are based (or where they spend their holidays), especially in south-east England where there are the greatest numbers of both species and people. However, given sufficient time, guidance from the scheme organiser, and adequate feedback to recorders, more regular coverage over much of the Britain can be obtained for most groups. Ireland is less well recorded because it has a smaller population with proportionally fewer specialists than Britain, but even there it is possible to motivate recorders and obtain adequate coverage in a decade or less (see Doogue and Harding 1982; Ronayne 1987).

(d) Knowledge of the biology of species

Knowledge of the biology of species is essential for efficient recording. Probably one of the most startling discoveries in the British flora was of early Star of Bethlehem *Gagea bohemica* which was originally found in late spring and thought to be a new population of Snowdon lily *Lloydia serotina*. Only when the new site was visited in early spring was its true identity discovered because it was in flower. Clements and Alexander (1987) have demonstrated how knowledge of larval biology has increased records of a snipe fly *Xylophagus ater*, previously thought to be nationally rare.

(e) Sampling techniques

Some species live in inaccessible places such as mountain tops or caves. Such species are difficult to record, but present a challenge to specialists who often go out of their way to record them. More difficult are species which occur within the general countryside, but in inaccessible habitats such as deep water, soil or interstices in gravels.

8.5.3 Status of populations

(a) Fugitive species

Some species, especially vascular plants, occur at a site for only a few years, often as pioneers of open or temporary habitats. One such example, grass poly (*Lythrum hyssopifolia*), now occurs very infrequently in the British Isles, but was always scarce (Preston and Whitehouse 1986). The fairy shrimp (*Chirocephalus diaphanus*) occurs in temporary pools widely throughout southern Britain (Bratton and Fryer 1990).

(b) Mixed native and introduced populations

Mixed populations are a problem which is commonly observed with vascular plants, for example, native black poplar (*Populus nigra*), the presence of which in Britain was confused with introduced hybrids although the native species was recognised as taxonomically distinct (Milne-Redhead 1990).

(c) Established aliens

The distribution of aliens established and reproducing in the British Isles can often seem eccentric. For example, the terrestrial amphipod (*Arcitalitrus dorrieni*) seems to have been accidentally introduced to south-west Britain (where it is spreading) and to a few isolated locations further north (where it has maintained populations, but not spread) (Harding and Sutton 1988). The interaction of alien species with closely allied native species (e.g. grey and red squirrels), or with species with which they may compete, predate or parasitise, provide many opportunities to use and interpret the results of distribution surveys. The ecology of introductions and invasive species was discussed by authors in Kornberg and Williamson (1987).

(d) Species of uncertain status

The origins of island floras and faunas, such as those of the British Isles, provide opportunities for research and interpretation. Evidence of the presence of vascular plants, beetles and molluscs is preserved as Flandrian sub-fossils, mainly in lowland peats and lake sediments, and at archaeological sites. Godwin (1975) and authors in Walker and West (1970) documented the origins and history of the British flora, but only limited interpretation of the invertebrate sub-fossil record has yet been attempted.

However, there is good, but rather diffuse, evidence that a substantial element of the British Isles fauna has been added by association with man (Kerney 1966 and authors in Sleeman *et al.* 1986).

8.5.4 Changes in the distribution of species

Using data collected by national surveys, including historical information from collections, herbaria and publications, it is possible to demonstrate changes in the distribution of species.

Many of the distribution changes can be linked to causal factors induced by man such as changing land use and land management leading to loss of suitable habitat for individual species (e.g. the silver-spotted skipper butterfly (*Hesperia comma*) Heath *et al.* 1984) or assemblages of species such as those of grassland (Wells and Sheail 1988) and of dead wood in forests (Speight 1989). Pollution is an often quoted cause of the decline of lichens such as *Parmelia caperata* and *Lobaria pulmonaria* (Seaward and Hitch 1982) and other organisms. Some species suffered badly from agricultural pesticide use (e.g. peregrine *Falco peregrinus*, Ratcliffe 1980).

However, the causes of the decline of species such as the nightjar (*Caprimulgus europaeus*) (Sharrock 1976) and the black-veined white butterfly (*Aporia crataegi*) (Pratt 1983) are less obvious. Similarly, the spread of the collared dove (*Streptopelia decaocto*) (Sharrock 1976) throughout Europe since 1930, or of the migrant hawker dragonfly (*Aeshna mixta*) (Merritt *et al.* in press) in the UK over the last 30 years cannot easily be explained.

Climate is a contributory factor in changes in the range of some species. Pollard (1979) demonstrated that the expansion of range of the white admiral butterfly (*Ladoga camilla*) in the 1930s and 1940s was related to warm summers. A similar expansion of the comma butterfly (*Polygonia c-album*), following a major contraction in range in the second half of the 19th century, is less clearly related to weather (Pratt 1986). In the general context of climate changes, national surveys of invertebrates, many species of which have a life cycle of only one year, are likely to produce evidence of changes in our wildlife long before changes can be discerned in vascular plants.

8.6 ROLE OF SURVEYS IN MONITORING

The distinctions between surveys, surveillance and monitoring have been defined in earlier chapters. National species distribution surveys have the potential to provide baselines for future monitoring of the flora and fauna of sites and grid cells. However, it is essential to be aware of the limitations of data from national surveys. Such data are extensive, whereas monitoring is normally intensive and localised. Data from surveys are not usually collected to provide a basis for monitoring and should be seen to augment the results of monitoring – putting the results in a wider context.

8.6.1 Surveys as baselines

Most national species distribution surveys collect site-relatable data with a precise temporal component. Thus, information on the presence of species at sites is documented to provide a baseline from which to assess and monitor future changes.

Other authors have described the need to monitor species and assemblages for a variety of purposes, for example to assess the effects of:

- land-use changes;
- land and water management (including wildlife conservation and recreation);
- wildlife conservation legislation;
- pollution;
- pesticide use;
- climate change.

The information and expertise that result from national species distribution surveys provide a resource to underpin monitoring needs by being able to relate the sample of monitored data to overall national data.

By their very nature, most national surveys are long-term projects – few have sufficient support and expertise to be able to completely survey the British Isles in less than a decade. Long-term observation and recording of species provide opportunities to assess the effects of changes in the natural environment and so validate models. Assessment and validation can be in two time scales – retrospective and in the future, from a baseline.

Retrospective assessment of changes is possible where information has been gathered over a period of decades, or where information can be gleaned from other sources such as museum or personal biological collections or published records. Most national species distribution surveys aim to collate information from such sources, to provide a historical perspective to their current data. In the British Isles, Heath *et al.* (1984) demonstrated the use of historical records to show the decline of many species of butterflies, and Geijskes and van Tol (1983) showed the loss of dragonflies in the Netherlands using data from collections.

Partial or total re-surveys of some taxonomic groups are already being conducted from baselines of earlier surveys.

1. The Botanical Society of the British Isles (BSBI) has completed its Monitoring Schemes, a partial re-survey (11% of 10 km squares with a selected network of 2×2 km squares) to be used in future to monitor

changes in the flora (Ellis 1986). These recent data from sample squares will also be compared with the results of the Atlas 10 km square survey (1954–9) to provide an objective assessment of the species which have changed in frequency over the last 30 years.

2. The British Trust for Ornithology is presently conducting a complete re-survey of breeding birds (Gibbons 1988) which will enable comparisons to be made with the earlier survey of breeding birds (Sharrock 1976).

3. The initial phase of a national survey provides the baseline and develops the expertise of volunteers. Having completed a national survey, these resources of data and expertise can be re-deployed to extend recording. Re-survey of selected important breeding sites for dragonflies, to assess breeding success, is the main objective of the Odonata Key Sites Project launched in 1988. Regular re-survey of selected grid cells or sites may be interpreted as monitoring on a long time-scale.

The application of multivariate techniques and the use of geographical information systems to species, other biotic and abiotic environmental datasets enable the products of surveys to be analysed and interpreted. Models to predict the consequences of environmental changes, such as climate and land use, are being constructed. The predictions of these models and the results of analyses should form a basis for future projects to monitor the real effects of changes on species and other assemblages.

8.6.2 The importance of long-term studies

The importance of long-term studies, for so long the Cinderella of scientific research, has recently come to be recognised by science administration. For example, the Department of the Environment commissioned a review of potential sources of data to compile statistics on key indicators of 'the health' of British wildlife (Crawford, Toy and Usher 1989). A broadly-based working party prepared a report (NERC 1989) on long-term reference sites in Britain which advocated the absorption of selected sites into a formalised national network of monitoring sites.

National species distribution surveys lie somewhat on the fringes of these two studies. However, recognition of the need for surveys, to provide baselines for further studies, as well as a national perspective to sample studies such as the Butterfly Monitoring Scheme (Chapter 6) or county-based monitoring (Chapter 9), was given in the report of a Linnean Society working party (Berry 1988).

A movement to emphasise the importance of biological recording has been gathering momentum in the 1980s. This movement, which is reviewed by Harding (1990), is coming to a peak of activity at a time when the value of long-term surveys and monitoring is being recognised, particularly in relation to climate and land-use changes.

The need for national policies and co-ordination in biological surveys and recording has been recognized at various levels in the scientific community in the UK, and internationally by the Council of Europe and the European Community. It is to be hoped that awareness of the need for such policies can be fostered and encouraged in central government as part of a general process of increasing environmental concern and enlightenment.

REFERENCES

Berry, R.J. (1988) *Biological survey: need and network*, PNL Press/Linnean Society of London, London.

Bratton, J.H. and Fryer, G. (1990) The distribution and ecology of *Chirocephalus diaphanus* Prévost (Brachiopoda: Anostraca) in Britain, *Journal of Natural History*, **24**, 955–64.

Clapham, A.R. (1951) A proposal for mapping the distribution of British vascular plants, in *The study of the distribution of British plants* (ed. J.E. Lousley), Botanical Society of the British Isles, Oxford.

Clements, D.K. and Alexander, K.N.A. (1987) The distribution of the fly *Xylophagus ater* Meigen (Diptera: Xylophagidae) in the British Isles, with some notes on its biology, in *Proceedings and transactions of the British entomological and natural history society*, **20**, 141–6.

Copp, C.J.T. and Harding, P.T. (1985) *Biological Recording Forum 1985*, Biology Curators' Group Special Report No 4. Biology Curators' Group, Bolton.

Crawford, T.J., Toy, R. and Usher, M.B. (1989) *Key indicators for British wildlife* (Stage 1). Unpublished report to the Department of the Environment. University of York, Institute for Applied Biology, York.

Crundwell, A.C. and Hill, M.O. (1977) *Anoectangium warburgii*, a new species of moss from the British Isles, *Journal of Bryology* **9**, 435–40.

Doogue, D. and Harding, P.T. (1982) *Distribution atlas of woodlice in Ireland*, An Foras Forbartha, Dublin.

Ellis, R.G. (1986) The new mapping scheme, *B.S.B.I. News.* **43**, 7–8.

Gibbons, D. (1988) The new atlas of breeding birds in Britain and Ireland, *BTO News,* **154**, 1–2.

Godwin, Sir H. (1975) *The history of the British flora. A factual basis for phytogeography* (2nd ed.), Cambridge University Press, Cambridge.

Geijskes, D.C. and van Tol, J. (1983) *De libellen van Nederland (Odonata)*, Koninklijke Nederlandse Natuurhistorische Vereniging, Hoogwoud.

Harding, P.T. (1985) *Biological Records Centre – a national data bank*, Biological Records Centre, Huntingdon.

Harding, P.T. (1989) *Current atlases of the flora and fauna of the British Isles*, Biological Records Centre, Huntingdon.

Harding, P.T. (1990) Biological survey: need and network – a review of progress towards national policies, in *National perspectives for biological recording in the United Kingdon*, (eds G. Stansfield and P.T. Harding), NFBR, Cambridge.

Harding, P.T. and Sutton, S.L. (1985) *Woodlice in Britain and Ireland: distribution and habitat*, Institute of Terrestrial Ecology, Huntingdon.

Harding, P.T. and Sutton, S.L. (1988) The spread of the terrestrial amphipod *Arcitalitrus dorrieni* in Britain and Ireland: watch this niche!, *Isopoda,* **2,** 7–10.

Heath, J. and Perring, F.H. (1978) *Biological Records Centre,* Institute of Terrestrial Ecology, Cambridge.

Heath, J., Pollard, E. and Thomas, J. (1984) *Atlas of butterflies in Britain and Ireland,* Viking, Harmondsworth.

Hopkin, S.P. and Roberts, A.W.P. (1987) A species of *Haplophthalmus* new to Britain, *Isopoda,* **1,** 37–48.

Kerney, M.P. (1966) Snails and man in Britain, *Journal of Conchology,* **26,** 3–14.

Kornberg, Sir H. and Williamson, M.H. (1987) *Quantitative aspects of the ecology of biological invasions,* The Royal Society, London.

Lack, P. (1986) *The atlas of wintering birds in Britain and Ireland,* Poyser, Calton.

Milne-Redhead, E. (1990) The B.S.B.I. black poplar survey, 1973–88, *Watsonia,* **18,** 1–5.

Merritt, R., Moore, N.W. and Eversham, B.C. (in press) *Atlas of the dragonflies (Odonata) of Britain and Ireland,* HMSO, London.

Nature Conservancy Council (1989) *Guidelines for selection of biological SSSIs,* Nature Conservancy Council, Peterborough.

NERC (1989) *Report of the working party on long term reference sites.* Unpublished report by the Natural Environment Research Council.

Perring, F.H. (1971) The Biological Records Centre – a data centre, *Biological Journal of the Linnean Society,* **3,** 237–43.

Perring, F.H. and Farrell, L. (1983) *British Red Data Books: 1 Vascular plants.* (2nd ed.) Royal Society for Nature Conservation, Lincoln.

Perring, F.H. and Walters, S.M. (1962) *Atlas of the British flora,* Nelson, London.

Pollard, E. (1979) Population ecology and change in range of the white admiral butterfly *Ladoga camilla* L. in England. *Ecological Entomology,* **4,** 61–74.

Praeger, R.L. (1901) *Irish topographical botany. Proceedings of the Royal Irish Academy, 3rd Series,* **7,** *clxxxviii* + 1–410.

Praeger, R.L. (1906) A simple method of representing geographical distribution, *Irish Naturalist,* **15,** 88–94.

Pratt, C. (1983) A modern view of the demise of *Aporia crataegi* L., the black-veined white, *Entomologist's Record Journal of Variation,* **95,** 45–52, 161–6, 232–7.

Pratt, C. (1986) A history and investigation into the fluctuations of *Polygonia c-album* L. the comma butterfly, *Entomologist's Record Journal of Variation,* **98,** 197–203, 244–50; **99,** 21–7, 69–80.

Preston, C.D. and Eversham, B.C. (in preparation) Recording bias and species mobility as factors affecting the interpretation of distribution maps.

Preston, C.D. and Whitehouse, H.L.K. (1986) The habitat of *Lythrum hyssopifolia* L. in Cambridgeshire, its only surviving English locality, *Biological Conservation,* **35,** 41–62.

Ratcliffe, D. (1980) *The Peregrine Falcon,* Poyser, Calton.

Ronayne, C. (1987) *Provisional distribution maps for Odonata in Ireland.* Irish Odonata Recording Scheme, Skerries. (Unpublished.)

Seaward, M.R.D. (1984, 1985) *Provisional atlas of the lichens of the British Isles: Vols 1 and 2.* School of Environmental Science, University of Bradford, Bradford.

Seaward, M.R.D. and Hitch, C.J.B. (1982) *Atlas of the lichens of the British Isles: Vol. 1.* Institute of Terrestrial Ecology, Cambridge.

Sharrock, J.T.R. (1976) *The atlas of breeding birds in Britain and Ireland*, Poyser, Calton.

Shirt, D.B. (ed.) (1987) *British Red Data Books: 2. Insects*, Nature Conservancy Council, Peterborough.

Sleeman, D.P., Devoy, R.J. and Woodman, P.C. (1986) *Proceedings of the Postglacial Colonization Conference, 1983*. Occasional publication of the Irish Biogeographical Society, No. 1. Irish Biogeographical Society, Dublin.

Speight, M.C.D. (1989). *Saproxylic invertebrates and their conservation*, Nature and Environment Series No. 42. Council of Europe, Strasbourg.

Stearn, W.T. (1973) *Facsimile version of John Ray: Synopsis methodica stirpium Britannicarium, editio tertia 1724; and Carl Linnaeus: Flora Anglica 1754 and 1759*, The Ray Society, London.

Walker, D. and West, R.G. (eds) (1970) *Studies in the vegetational history of the British Isles*, Cambridge University Press, Cambridge.

Watson, H.C. (1859) *Cybele Britannica or British plants and their geographical relations. Vol. 4*, Longman, London.

Watson, H.C. (1873–4) *Topographical botany, parts 1 and 2*. Thames Ditton (privately published).

Wells, T.C.E. and Sheail, J. (1988) The effects of agricultural change on the wildlife interest of lowland grasslands, in *Environmental Management in Agriculture, European Perspectives* (ed. R. Park), Belhaven, London, pp. 186–201.

– 9

Monitoring at the county level

CLAIRE E. APPLEBY

Monitoring can be defined as 'biological recording with the specific aim of detecting changes in the distribution and abundance of species and their habitats' (Radford 1986). At county level this is carried out by local organisations such as the county wildlife trust or the local biological records centres (LBRCs). With a wide range of functions and limited resources these bodies must adopt a very pragmatic approach to monitoring – the rigours of an academic discipline are a luxury they cannot afford.

This chapter will cover the role of LBRCs, describe a suitable computer system and suggest how the county centres can function as a national network. Throughout, the emphasis will be on techniques suitable for monitoring at county level and will embrace monitoring of species, sites, habitats, land-use change and other factors.

9.1 ROLE OF COUNTY BIOLOGICAL RECORDS CENTRES

A county biological records centre exists to provide information on the wildlife and environment of the county to whoever may need it. Appropriate data exist in a wide variety of forms; one of the centre's main functions is the collation of these data, and many of its other activities result directly from this role. The centre acts as a link between the recorders and the users (Figure 9.1). It must bring the data together in a way which allows them to be interrelated, and it must adapt their presentation to the needs of the user. It is the fact that this can be achieved, given appropriate technology, that is one of the local records centre's greatest strengths.

Unlike previous chapters where monitoring may have been restricted to a particular technique (e.g. remote sensing) or a taxonomic group (e.g. butterflies), for a county centre all techniques and all species must be considered – the only restriction is the geographic extent of the data.

The great variety of techniques required can be demonstrated by looking at the many uses to which the data are put. The LBRCs most conspicuous product is probably the atlas of county flora, a compilation of distribution maps of individual species. However, the LBRCs role goes much wider. Uses of the data can be broadly classified as planning, land use, conservation,

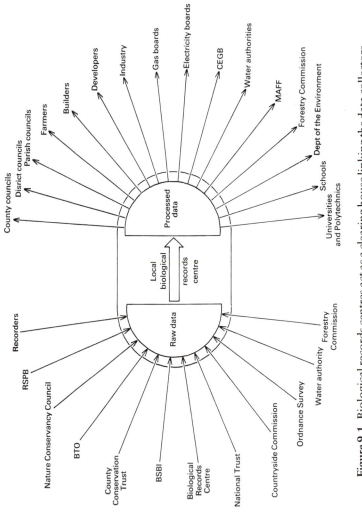

Figure 9.1 Biological records centres act as a clearing house linking the data collectors with those who use environmental data.

ecology and education, although the classification is, of course, artificial and there is much overlap between classes.

9.1.1 Planning

The county records centre often works closely with the planning department of the county and district councils. The planning structure is hierarchical, ranging from individual planning applications, through parish and district planning strategies to the full county structure plan. The LBRC will contribute at all levels.

Individual planning applications need to be processed rapidly and the centre will probably monitor the weekly lists sent out by the district councils. A development might be proposed at any site within the county and the LBRC must therefore endeavour to hold up-to-date information at least for every greenfield site in the county. This weekly monitoring is time-consuming but the LBRC may, through co-operation with individual parishes, be able to devolve some of this work. The first step is the preparation of a parish appraisal, by the parishioners, using information supplied by the LBRC and supplemented by data gathered in the field. For more information on preparing parish appraisals and parish maps see King and Clifford (1985).

Where major developments are proposed the LBRC may be able to contribute significantly to an environmental impact appraisal, providing data local to the site as well as placing the site in its county, regional, or even national context. The centre's detailed, local information must then be summarised in a form suitable for district strategies and for the county's structure and subject plans. The use of environmental data in planning is described in more detail by Francis (1984) and Evans (1986).

9.1.2 Land use

The LBRCs contribution to planning will include some advice on land use, but not all forms of land use come under development control. Wherever possible the LBRC should contribute to the development of land-use strategies concerning agriculture and forestry and should provide advice to water authorities and other utilities. In all cases, the earlier the LBRC is consulted, the more influence it is likely to have.

9.1.3 Ecology

The LBRC has an extensive role to play in the related fields of conservation, ecology and general natural history. Its work falls into three categories: the simple provision of data, the analysis of the data to provide new information, and the support of the large network of amateur naturalists. The extent to

(a) Enallagma cyathigerum

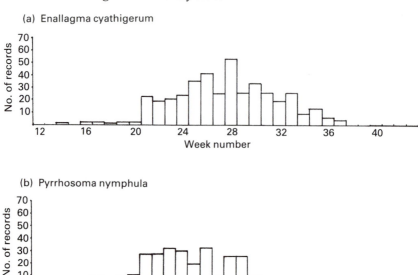

(b) Pyrrhosoma nymphula

Figure 9.2 Comparison of dragonfly flight periods through analysis of recording dates.

which LBRCs fulfil these roles depends on the resources, expertise and inclinations of individual county centres.

Even the simple presentation of data, with neither interpretation nor analysis, has the potential for almost limitless variety, discussed further in section 9.4.4. However, existing data can be analysed to show inherent, but not apparent, patterns. For example, analysing recording dates can demonstrate seasonality of species (from day and month of record) (Figure 9.2) or the spread or decline of species (using the year of record). Interpretation of data can be provided by, for example, collating data for individual sites, and, by comparison with other sites, providing an evaluation for conservation purposes.

Finally, the LBRC must nourish its indispensable network of amatuer recorders, providing advice and encouragement, giving guidance when required on the organisation of surveys, and providing constant feedback through newsletters, reports, published atlases and so on.

9.2 DATA

The centre is not just concerned with species distribution, but must consider all factors that might affect the ecology of a species. The range of data is therefore as follows:

(a) *Species*

For species information the records centre needs the minimum of species name, recorder, date and location and may also hold details of populations, age, determiner, voucher specimen and so on. The records might be presence/absence data from recording schemes or one-off records, or site-related data from site surveys which might be quantitative or presence/absence.

From this information the centre can produce the familiar and important distribution maps (Figure 9.3).

(b) *Habitat*

The centre should hold information about all the important habitats across the county. This may largely take the form of detailed information about

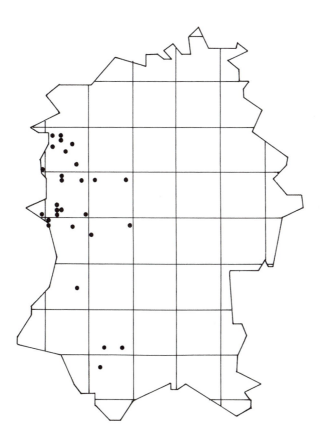

Figure 9.3 Distribution of *Helleborus foetidus* in Wiltshire.

particular sites derived from phase 2 surveys* and special surveys such as the Inventory of Ancient Woodland, Sites of Special Scientific Interest or the Invertebrate Site Register. It should include information such as conservation status, management history, access and ownership.

(c) Environmental factors

General information will be required for the whole of the county and can be obtained from the Soil Survey, the Geological Survey, the Ordnance Survey and from field survey.

Such information would include:

- land use
- vegetation cover
- phase 1 survey data
- landscape
- geology
- soils
- altitude
- slope
- aspect

When seeking information on these features, the records centre should aim for complete county coverage.

9.3 RECORDING PROCEDURES

(a) Data collection

Techniques for data recording need to be standardised in order to ensure comparability of records. LBRCs therefore need to establish clear guidelines for the collection of data. County centres collect or promote the collection of biological and environmental data. Resources rarely enable them to gather data in the field but a wealth of information already exists, lodged with a variety of county and national organisations, public bodies and individuals. The centre's reputation depends on the quality and reliability of its data; it should ensure the preservation of voucher material where appropriate and it should identify gaps in taxonomic or geographic coverage and actively promote recording in these areas. For a detailed account of validation see Irwin (1985).

(b) Integration

The collation of data is crucial to the centre's role as a clearing house. Each centre must endeavour to trace all existing information for the county. Much

* For clear accounts of NCC's three-phase method of surveying see Bines (1985) and Keymer (1986).

of this information will be held by organisations located outside the county. All biological and environmental data must then be integrated in order to maximise their use. Having been derived from a wide range of sources, the information will be in a variety of different forms, so that integration is by no means a simple task. The key here is that most data have a spatial element, that is they are related to a particular location in the county, whether a grid reference, a site or some other unit. This spatial element can be used as the base referencing system so that each type of data can be mapped and related to other data in the manner of a map overlay.

(c) Use of data

The centre will need to analyse, correlate and otherwise interpret the data, both for its own use and to meet the needs of other users. To increase its effectiveness the centre should actively promote the use of the data, and it must be able to disseminate these data in a wide variety of forms.

9.4 TECHNOLOGY

The records centre acts as a focal point, matching the wide range of data with the equally wide range of users. The variety of data and functions encompassed by the centre are such that choice of the right computer system becomes extremely important. Only by the use of the appropriate technology can a records centre achieve its full potential and the only way to accomplish this task with limited manpower is by using a well-designed computer system which will allow survey data in any form to be stored, and output in a totally different form.

Computerisation is essential for it is the only way to cope with the almost unlimited amount of information and the severely limited resources of most records centres and county trusts. However, the computer must not be seen simply as a means of carrying out more rapidly those tasks that were previously performed manually, and often rather laboriously. The computer also permits many new operations which were impractical if not impossible before.

In choosing a computer system it is important to recognise that almost all environmental data have a spatial element and that it is frequently the spatial relationships between data that are important. It is therefore essential to adopt a system that can handle these spatial features in addition to carrying out normal database functions. Also, from the users' viewpoint, most planners deal with maps, most members of the public are more receptive to visual patterns than to text, and however misguided it may be, many people judge information by the quality of its presentation. A costly Geographical Information System is probably the ideal solution but, for a much lower cost a good graphics package linked to a database will give excellent results.

The main features of any system are:

- input
- manipulation, and
- output

However, the techniques used depend on the way in which the data are stored so this will be dealt with first.

9.4.1 Data storage

In general terms the system will use one of two methods for data storage.

1. The database will store textual information, mostly in the form of codes, as a standard sequential file containing individual records constructed from a number of uniform fields. For example, Figure 9.4 shows 12 records (numbers 131 to 142) from a database which includes a five-character field containing species code, an eight-character field for grid references and a date field.

2. The graphics package will store information related to co-ordinates based on the national grid, as either a single point or as a string of co-ordinates defining a line or boundary on a map. Each feature is given a separate layer, equivalent to a transparent overlay, and layers can be turned on and off to display whatever combination is required. For example, in Figure 9.5 a species record has been combined with the course of a river and its tributaries, with the boundary of a nature reserve and with the boundary

RECNO	SPECIES	GRIDREF	DATE
131	10135	SK109583	23-06-87
132	10135	SK345983	12-12-56
133	10135	SK912847	01-04-86
134	11376	SK659798	28-07-82
135	16544	SK783489	21-03-65
136	16728	SK798349	30-08-76
137	07346	SK099878	02-09-74
138	10045	SK657978	21-10-88
139	12868	SK788934	30-09-55
140	10045	SK284683	22-12-45
141	10045	SK985341	12-06-86
142	16728	SK784502	01-11-67

Figure 9.4 Example of a simple database file.

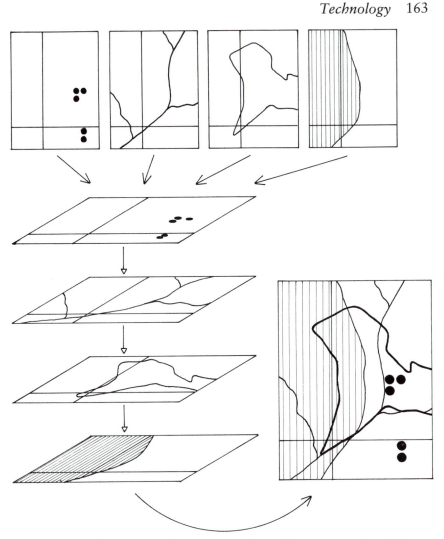

Figure 9.5 Use of computerised overlays to combine different types of spatial data.

of an AONB. Note how each additional piece of information helps to place the other information in context.

The combination of database and graphics gives the records centre the means to store all its information in the best possible way. The three types of information discussed earlier would be computerised as follows.

1. Species records would form a conventional database file but the records would be transferred to the graphics package to give a distribution map.

This map can be superimposed on other graphic elements such as soils to show how the distribution relates to environmental factors.

2. The boundaries of nature reserves and other sites would be digitised together with the distribution of habitats and other features within the site. Other information such as management details and conservation status would be held in a site file in the database.

3. Most countywide features would be stored as digitised boundaries showing the extent of each feature.

9.4.2 Data input

It is most important for data to be input as quickly as possible to enable the records centre to build a comprehensive database which is continually updated. This must be achieved, however, without compromising accuracy. Wherever possible, information should be transferred to the records centre in machine-readable form, either on disk or over a network or telephone line.

Keyboard entry of data can be time-consuming and should be reduced to a minimum. However, with a fast typist and using carefully designed data entry programmes, speeds of, say, 100 000 records per man/month can be achieved. It is better to design a screen entry form which will accept up to 25 records rather than having to refresh the screen after every record. Computerised validation of records is also time-consuming and should not be incorporated into the data entry programmes but should be run later as a batch programme. Every second is important since reducing data entry time by one second per record saves about eight weeks work on a million records.

Information from maps can be entered using a digitising tablet. The map is placed on the tablet and a line on the map is traced using an electronic pointer. The tablet registers the position of the pointer and converts this information to a string of co-ordinates which can be used to reproduce the line. Information is digitised once only but can then be presented at a whole range of different scales so that maps of parishes or even small sites are as easy to produce as maps of the whole county.

9.4.3 Data manipulation

Once the data are computerised the different data sets need to be merged. The overlaying facility, briefly mentioned above, and the interaction between the graphics and the database enable survey data of many different types to be stored together in a single system and give the records centre great flexibility. Elements of different surveys can then be combined for analysis, interpretation and display according to the users' requirements.

The database and the graphics system manipulate the data in different ways and will be dealt with separately.

(a) Database functions

Database structure has been outlined above. The strength of the database system is its ability to search and sort on a variety of fields, or combinations of fields, and to carry out mathematical functions using the data.

Searching   From a file of a million records or more a database package must be able to rapidly select those records relevant to a particular query. The operator structures the query using a predefined syntax and the computer searches the appropriate database fields, comparing their values with the values set out in the query. For example, if the operator wanted all records of *Heracleum mantegazzianum* between 1950 and 1980, the computer would search for an exact match in the species field and any date within the specified range.

A good database package would normally have the ability to select all records containing the following:

1. an exact match on one field, e.g. all records of *Conopodium majus*;
2. an exact match on two or more fields (all records of *Gryllotalpusa gryllotalpa* from Kent in 1984);
3. a match on one field but excluding records matching on a second field (records from Salisbury Plain excluding Red Data Book species);
4. search for a minimum number of a list of attributes (sites with at least seven out of ten ancient woodland indicator species);
5. search for the absence of a particular attribute or attributes (species for which there are no records after 1950);
6. count the number of records containing particular attributes (the number of sites in the county containing a Red Data Book species);
7. cross-reference via common fields (use the species code to insert English and/or scientific names into a list of species records).

Sorting   When preparing reports which contain long lists of records the ability of the computer to sort the records is important. For example, species records may be presented in alphabetical order of Latin name or in taxonomic order. Within this sequence the records may be sorted by year or by grid reference. Very long lists may require quite complex sorting procedures. For example, a list of all plant records for the whole county may be sorted first by species, then by 10-km grid squares, then alphabetically by locality within each square, then according to date and so on. Organising the records in this way helps to make long lists more comprehensible.

Mathematical functions   While the search produces a group of records, the computer can also carry out mathematical comparisons and calculations within the group or between groups. Thus it is possible to select the largest site on which a particular species occurs or the earliest and latest records for one species. It is also possible to calculate the number of records of one species or the average number of species per 10-km square. It is only a short step from this to carrying out simple statistical comparisons on the data, or to formatting the data for transfer to more complex specialist statistical packages.

(b) Graphics facilities

Manipulation of the graphics data uses a completely different range of facilities to the database work.

Changes of scale   The graphics package allows maps to be viewed on the screen or plotted on to paper. When on the screen, any section of the map can be defined by a 'window' or rectangular frame and this portion can be expanded to fill the screen. This is one of a number of methods that can be used to view maps at any scale. When plotting on paper, a window can be chosen to fill a stated size of paper or a precise scale such as 1 : 20 000 can be specified. In this way the records centre can produce maps ranging from a general map of the whole county through parish maps to a detailed map of a particular field.

Layers   As mentioned above, the graphics information is organised into layers, equivalent to transparent map overlays, so that each layer represents one particular feature. For example, chalk grassland may be held on one layer, SSSIs on another, while other layers may store parish boundaries, contours, rivers, AONBs and so on. All layers can be turned on or off so that features can be displayed in any combination. When a map is plotted on paper only those layers turned on will be included. This facility has many uses: when producing a map in answer to a query only the most relevant layers need be included; very detailed information on individual sites can be confined to layers which are never used when a county perspective is required; and when information is altered, for example when an SSSI boundary changes, the old data can be kept on a separate layer to give a historical perspective.

Measurement   For any digitised line the computer can calculate the length of the line or the area of land bounded by that line. Thus lengths of river stretches, sizes of SSSIs or areas of chalk grassland can be obtained without difficulty. The software will also provide a great deal of statistical information such as the number of occurrences of a particular symbol which might represent the location of a spring, for example, or a tree preservation order.

Landscape   Many graphics packages can use contours to produce a three-dimensional image (Figure 9.6). The more sophisticated can 'drape' this image with features such as habitat or land use and can show the viewpoints from which a structure of given height would be visible. Thus the visual impact of development or other land-use change can be assessed.

(c) Graphics/database interaction

The two main components of the computer system each have powerful facilities for manipulating certain types of data (either text or graphics). Further flexibility comes from the ability to exchange data between the two components.

Database to graphics   The database can return a list of 10-km squares in which a particular species has been recorded and from these data the graphics package can locate symbols on a species distribution map. Different symbols might be used to indicate, for example, post-1970 and pre-1970 records or

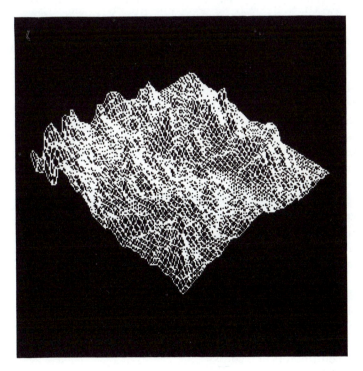

Figure 9.6 Digital terain modelling – use of height data to produce a three-dimensional image.

different sized symbols might be used to show the number of records per square or the precision of the record.

Graphics to database   Information obtained by the graphics system, particularly that described in section 9.4.3.2c, can be transferred to appropriate records in the database for rapid retrieval later on. Individual graphics items can be labelled in a predefined way. For example, sites may be labelled with their name, the survey from which the information was obtained and the date of the survey; the type of site (e.g. SSSI) is determined by the graphics layer. Using this information the database could extract names of all nature reserves surveyed before 1960.

9.4.4 Output

Output must be very flexible to cater for the widely varying needs of different users. Maps, lists, tables, statistics and narrative reports must be combined in a presentation which is cohesive and easy to understand. The examples listed below give some idea of the wide range of output formats required.

The following may be requested in relation to a particular site:

- site map
- location map (position in county)
- narrative description
- number of plant and animal species
- list of people who have contributed records
- full species list in alphabetical order of English names
- full species list in taxonomic order
- list of plants only, in alphabetical order of Latin names
- previous casework
- general data summary
- full detailed information including references to written sources and to surveys.

The following may be requested in relation to a particular species:

- national or international conservation status
- county distribution map
- parish distribution map
- pre-1915 distribution map
- list of records from specified five-year period
- number of 1 km squares from which species has been recorded
- list of contributors of records
- graph showing number of records of species plotted against, for example, altitude

- statistical analysis of relationship plotted above
- analysis of distribution of plant species in relation to, for example, soil type.

The following may be requested in relation to several species:

- a map showing the occurrence of two species together
- statistical analysis of the coincidence of two species
- list of species recorded from five or fewer 1 km squares
- list of Red Data Book species
- number of species plotted against land-use categories
- map showing the number of species recorded for each hectare square.

The following may be requested in relation to habitats:

- map showing rivers in the county
- map of one river system
- total length of rivers in that system
- length of river in each water quality class
- map showing distribution of chalk grassland
- number of chalk grassland sites, total area and area of each site
- table listing all woods in the county with areas, dominant species and soil type
- table showing number of woods, and total area, on each soil type
- map showing habitats within one mile of proposed bypass route.

Other enquiries:

- distribution of SSSIs in the county
- distribution of SSSIs on chalk
- list of sites within one mile of proposed route of new pipeline
- map showing land ownership boundaries.

The form of output is dependent on the use to which the information will be put and this will be dealt with in detail in the next section.

9.5 USE OF THE DATA

The main purpose of collecting and storing environmental data is that it should be available in the most useful form to those who need it. With such a broad range of information the records centre is equipped to contribute to many important issues in the county. These can be grouped under five general headings:

1. Planning might include:
     strategic planning
     development control

Environmental Impact Analysis
parish appraisals and parish maps
2. Land use, for example:
   agricultural policy and practice
   water authorities policy and practice
   land management
3. Conservation:
   site protection
   habitat management
   identification and evaluation of sites
   monitoring of species and habitats
   water quality monitoring
   species status assessment
4. Natural history and ecology:
   support and guidance for amateur naturalists
   support of data collecting network
   preparation of local floras and faunas
   provision of general information service
   provide a local contribution to national surveys
   co-ordination with the national network
   research – pure and applied ecology
5. Education:
   schools from primary level to GCSE
   adult education
   interpretive publications and displays

A selective review of the uses of biological data can be found in Stansfield and Harding (1990).

The following examples of monitoring illustrate how the form of output can be varied according to the needs of the user. The monitoring process might be described simplistically as comprising four actions:

1. identify and record as much as possible to provide a base line for future monitoring;
2. observe change over time to provide models on which to base future prediction;
3. comment on the potential impact of climatic or land-use change;
4. monitor and record actual change occurring following climatic or land-use change.

9.5.1 Species distribution

Species records plotted in conventional form in national atlases give a distribution map where each symbol represents at least one record for the

10-km square in which the symbol occurs. County centres will usually use one symbol per 1 km square, otherwise data are lost. However, many records will not be this precise. Symbols of mixed size can be used on one map to indicate recording precision. Alternatively, symbols of different sizes may be used to indicate the number of records per square.

Having examined the computer system in some detail it will be seen that the production of distribution maps for one species for time periods of, say, ten years per map is a fairly routine operation. Analysis of these patterns is dealt with in detail in Chapter 8, but changes in distribution may reflect habitat, land use or climatic changes, competition from other (possibly introduced) species or even change in recording effort.

An indication of fruitful areas for further research can be obtained by plotting species distribution together with environmental information such as soils, geology or altitude. Relationships suggested by the visual patterns produced can be confirmed by careful statistical tests, but the speed with which different combinations of factors can be mapped allows rapid preliminary evaluation of many different relationships.

9.5.2 Habitat distribution

Changes in habitat distribution can be assessed visually by overlaying field survey maps from different years. Where insufficient habitat data are available changes in distribution of indicator species can be used as a guide to habitat changes. For example, a number of dragonfly species may be used as indicators of slow-flowing water, and their distribution can be contrasted with that of selected aquatic plants from swift-flowing streams (Figure 9.7). The use of indicator or constant species can give approximate distributions of important habitats such as ancient woodland, species-rich meadows or individual National Vegetation Classification classes. The comparison of statistical tests with these visual techniques provides confirmation, or otherwise, of the patterns shown.

9.5.3 Other examples

Pollution-sensitive species can be used for monitoring air and water quality. Again, distribution maps of, for example, lichen representing sequential time periods are used. A combination of land-use surveys, habitat surveys, information from aerial photographs and other sources can provide a picture of land-use change. This can be built on by recording results of planning applications, road construction, wood clearance and other changes as they occur. Species records can be used to show seasonality of flight periods, breeding or migration where changes may reflect climatic or other trends. Species records can also provide an indication of species diversity across the

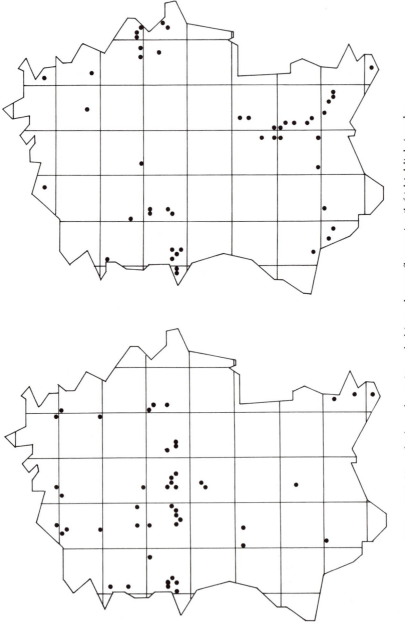

Figure 9.7 Use of selected species as habitat: dragonfly species (left) highlighting the course of a canal and various ponds, and selected aquatic plants (right) showing the location of fast-flowing rivers.

county which might be displayed on a map as the number of species per 10 km square. Changes in diversity might reflect land-use changes or other factors.

9.6 NETWORK

A county biological records centre does not always work in isolation from other such centres around the country or from other environmental bodies both local and national. An effective LBRC provides the data that its users require and if these relate to, for example, a river system they may not be confined to one particular county. It is frustrating for a researcher to find that the data available for one part of a river catchment are not uniform with those available for the rest of the catchment where this lies in another county.

It is therefore important for LBRCs to agree amongst themselves certain standard formats for collecting and storing data. For example, it is helpful if different centres can use the same habitat classification, the same taxonomic systems and so on. It is also useful if LBRCs can undertake to transfer to the appropriate county any data they receive which relates to that county, and if this can be done directly, computer to computer, this saves time and labour. It thus becomes necessary to lay down a standard format for information exchange.

Such co-ordination should exist also between LBRCs and national bodies, and between national bodies themselves. This would avoid organisations occasionally duplicating the work of others and thus conserve scarce resources in the environmental field. Co-ordination in all these areas would facilitate a smooth-running, more effective environmental movement.

9.6.1 County network

Much of a county biological records centre's work is related to planning, which has a hierarchical structure. The records centre will be directly involved at both county and district level, but it should not neglect the parish or 'grass roots' level. The centres rely for much of their data gathering on local volunteers who work close to their own homes. They can contribute records of flora and fauna to existing recording schemes, as well as provide information on land-use changes occurring within the parish; they can monitor planning proposals and alert the records centre to those with some environmental impact; and they can register their own evaluation of features within the parish by preparing a parish appraisal or parish map. The records centre can provide information and encouragement for the preparation of parish maps, although many of the features included will be historical, social or aesthetic. The parishes in turn can provide feedback to the LBRC on the accuracy of its data, they can supply additional information and can notify the centre of any changes. Contact can be maintained through an 'environmental watchdog' appointed in each parish. It is probably unrealistic to expect complete county coverage through this scheme but nevertheless a useful monitoring network can be established. Copp (1984) suggests that all LBRCs should define their 'Local Environmental Network', that is they should list all the activities related to the environment and recording and should note who is fulfilling each function in their county. Such a review would highlight areas of duplication or neglect and could lead to more formal relationships between organisations.

9.6.2 National components

At present, LBRCs are only of limited value beyond the county boundary. Lack of co-ordination means that there is considerable variation in administration, funding, recording methods, ease of access and so on. There is also no means of compiling data from each county to produce national statistics and there is no national database. However, the amount of existing data distributed amongst different organisations makes it impractical for any one body to attempt to hold it all.

There is a national network in the specialised area of species distribution. This network comprises the national Biological Records Centre at Monks Wood in Cambridgeshire and the numerous county BRCs. The network does not operate well because it is severely under-resourced and there is insufficient communication between its parts, but it has potential which may be wasted on the limited field within which it operates.

The Biological Records Centre is now part of a larger unit, the Environmental Information Centre, which embraces remote sensing, cartography and spatial data (see Radford, 1986). This unit is part of the Institute of Terrestrial Ecology. There are other relevant bodies such as the Geological Survey and the Freshwater Biological Association and NERC supports research in many relevant fields at universities and polytechnics. Then there is the Nature Conservancy Council, the Countryside Commission, the Department of the Environment, the Rural Areas Database at Essex University, the National Trust, the Royal Society for the Protection of Birds, the British Trust for Ornithology, the Botanical Society of the British Isles, the Soil Survey and many others.

9.6.3 Making it work

The present situation can, for convenience, be considered as a number of specific problems: lack of co-ordination; lack of statutory support; inadequate funding; insufficient attention to users' needs; failure to exploit new technology. Naturally, each of these problems affects some or all of the others and although it is helpful to consider each in turn, an effective solution must address all five in an integrated approach.

(a) Lack of co-ordination

The spread of data of various types among so many organisations brings a number of problems for both user and data gatherer. Much information may be 'hidden' to the user because it is difficult to discover who holds which data (see, for example, Copp, 1986). There is no central body to guide users to the organisation appropriate to the type of data they are seeking. Lack of communication between those gathering data can result in two or more organisations undertaking projects with very similar methods and aims. Duplication of effort results when a joint project could have saved valuable resources. There are no defined data standards and the combination of similar information from different organisations, or even within organisations, may be impossible due to inconsistencies such as different recording techniques or use of different habitat classifications. This is of particular relevance to monitoring; the NCC (1984) stresses that 'since monitoring is repeated survey, it is important that, whenever possible, initial surveys should be conducted in a standardised and repeatable manner'. These problems are compounded by the overall lack of resources. The importance of effective systems co-ordinated at national level, and the need for increased discussion at both national and regional levels, is frequently obscured by concentration of individual bodies on their own needs, in which the complexity of developmental problems and the urgency of establishing an effective system leaves little time for consideration of what other bodies are doing (Appleby 1985).

A co-ordinating body could provide standards and guidelines in a number of areas and could maintain a register of all those collecting and holding data on relevant subjects. The need for such a body was recognised in 1985 when about a hundred delegates met for a Biological Recording Forum at Chelsea College (see Stansfield 1985). The Forum subsequently led to the formation of the National Federation for Biological Recording which held its inaugural conference in 1986. The conference chairman reported that the time was right 'for a determined attempt to set up a properly organised and financed system for biological recording' (Stansfield 1986). Since then a working party set up by the Linnean Society has made similar recommendations (Berry 1988) and a co-ordinating commission has been charged, by the Natural Environment Research Council to examine the problem. The discussions continue, yet still the necessary resources elude both the Federation and the recording community.

(b) Lack of statutory support

It could well be argued that the co-ordinating role should be undertaken by one of the statutory bodies; in particular, the Department of the

Environment, Nature Conservancy Council (NCC) or National Environment Research Council (NERC) would appear suitable. The NCC emphasised the contrast between biological recording on the one hand and geological, soil and climatological recording to the other, when it resolved to 'urge government to accept recording of wildlife and human impacts as part of national environmental resource stock-taking' (NCC 1984). Certainly it would seem that some assistance should be forthcoming if only in the form of finance to enable a major review to be undertaken. Finance is a problem also at the local level, for most bodies devolve responsibility, for biological recording at least, on the NCC, even where the records centre encompasses the full range of environmental data. Unfortunately, the NCC maintained until recently a policy of not funding biological records centres. It may be that the only solution is to introduce legislation which obliges both statutory and other bodies to support recording at all levels.

(c) *Inadequate funding*
The problem of funding is crucial because although records centres and similar bodies can operate on a fairly modest annual budget, they do need sufficient resources for the following minimum facilities:

1. at least one full-time, permanent member of staff;
2. adequate computing facilities;
3. basic library of books and maps; and
4. since communication and co-ordination are crucial, adequate travelling budget for conferences and informal meetings.

The problem of obtaining funding from statutory bodies seems likely to continue unless major policy changes are made. However, another potential source of finance is the users of the data and this raises many problems of who to charge, how much and so on. It is also only a realistic option if the data suppliers begin to pay more attention to the users' requirements. One possibility for funding for a co-ordinating body is the raising of subscriptions from all those engaged in biological recording, in return for clearly defined services such as provision of data standards, advice on copyright, on the data protection act, and provision of a contact point or clearing house between suppliers and users.

County records centres have a responsibility to the national network. The immediate availability of accurate, up-to-date information is vital if the nature conservation movement in the UK is to respond to requests for advice, threats to sites, species and habitats, and routine considerations such as planning applications. Local centres therefore have a duty to:

1. contribute to the national network and accord with standards set nationally;
2. liaise with others, e.g. national bodies, local natural history societies, other LBRCs;

3. participate in national projects;
4. implement national policies and practices;
5. exchange experiences and ideas with other LBRCs throughout the country.

There is also a reciprocal responsibility for national organisations to communicate national developments to local records centres and to feed data from national surveys back to records centres local to the site of collection.

(d) Insufficient attention to users' needs

The records centre's role has been described in section 9.1 and the uses of the data explained in section 9.5. Those involved in environmental data supply can expect no support – statutory, financial or otherwise – if they do not provide a useful service. It is necessary to build an effective and efficient system based on a thorough understanding of users' needs, and to promote it to those who need it. In all probability it will then prove to be indispensable and will have no difficulty securing the support that it needs.

(e) Failure to exploit new technology

Naturalists and conservationists have been slow to exploit new technology. The NCC recognises that slowness to establish a computerised conservation database 'must be accounted a failure within the conservation movement as a whole, but of NC/NCC in particular' (NCC, 1984), while Appleby (1985) suggests that 'lack of resources, of readily accessible specialist advice, and also, perhaps, of imagination, may have led to opportunities of considerable potential being missed'. The importance of adopting systems that can store and manipulate spatial features has been discussed (section 9.4). Computerised systems of this type are available but are not utilised to any great extent in the biological field.

Identification of these problem areas must be followed by detailed examination leading to effective solutions, but these are tasks to which all those concerned, both users and suppliers, must together address themselves. However, the following steps towards a solution can be noted here.

1. A thorough review of users' needs must be undertaken so that decisions are made with full understanding of these needs, both existing and potential.
2. The full range of environmental data must be integrated.
3. Clear responsibilities must be established within the various statutory and non-statutory bodies.
4. A firm financial basis for environmental data supply must be secured.
5. A firm legal basis must also be established.
6. Effective lines of communication must be established both between suppliers and between users and suppliers.

7. The potential of new technology for efficient environmental data handling and supply must be fully investigated; in particular, the role of geographic information systems must be looked at.

Eventually an overall strategy must be developed, addressing all the problems discussed above and encompassing the whole range of suppliers from purely local to national or even international.

ACKNOWLEDGEMENTS

I would like to thank the Biological Records Centre for the data used to prepare Figures 9.2, 9.3 and 9.7, and the Environmental Information Centre for the diagram in Figure 9.6.

REFERENCES

Appleby, C.E. (1985) *Biological Recording: Computers and Conservation*, MSc thesis, University College, London.

Berry, R.J. (1988) *Biological survey: need and network*, PNL Press, London.

Bines, T.J. (1985) The work of the Nature Conservancy Council: a) habitat recording, *BCG Newsletter Supplement*, **4**(2), 15–18.

Copp, C.J.T. (1984) Local records centres and environmental recording – where do we go from here?, *BCG Newsletter*, **3**(9), 489–97.

Copp, C.J.T. (1986) The "hidden" data, in *Biological recording in a changing landscape* (eds P.T. Harding and D.A. Roberts), NFBR, Cambridge.

Evans, I.M. (1986) Involvement in planning, in *Biological recording in a changing landscape* (eds P.T. Harding and D.A. Roberts), NFBR, Cambridge.

Francis, K. (1984) The need of local authorities for environmental information. *BCG Newsletter*, **3**(9), 498–503.

Irwin, T. (1985) Validation of records, in *Biological recording forum* (eds C.J.T. Copp and P.T. Harding), BCG, Bolton.

Keymer, R.J. (1986) Survey and monitoring in the Nature Conservancy Council, in *Biological recording in a changing landscape* (eds P.T. Harding and D.A. Roberts), NFBR, Cambridge.

King, A. and Clifford, S. (1985) *Holding your ground*, Maurice Temple Smith, London.

Nature Conservancy Council (1984) *Nature conservation in Great Britain*, NCC, Shrewsbury.

Radford, G.T. (1986) An ecological data unit, in *Biological recording in a changing landscape* (eds P.T. Harding and D.A. Roberts), NFBR, Cambridge.

Stansfield, G. (1985) The problems being faced, in *Biological recording forum* (eds C.J.T. Copp and P.T. Harding), BCG, Bolton.

Stansfield, G. (1986) Chairman's introduction, in *Biological recording in a changing landscape* (eds P.T. Harding and D.A. Roberts), NFBR, Cambridge.

Stansfield, G. and Harding, P.T. (1990) *Biological recording: the products*, NFBR, Cambridge.

Field monitoring: confessions of an addict

KEVIN A. ROBERTS

10.1 INTRODUCTION

I suspect that most monitoring is thoughtless, so this chapter starts with some thoughts on the nature of monitoring. Monitoring is not as simple as collecting 'the facts'; nor do the facts 'speak for themselves': and *never* are interpretations 'not open to serious question', as one hopeful author wrote (Allison *et al.* 1952).

The second part consists of some personal monitoring tales. These derive from my work as a nature reserve warden, from expeditions abroad, from research to support the Lee Valley Conservation Group, and from studies with my students of the University of London's Certificate in Ecology and Conservation. I found the practical experiences invaluable at the time because they demonstrated a lot of strange things about monitoring which I wish I had thought of before.

Knowing what you are recording is crucial – but identification is not easy if you cannot see them (see rain-forest birds), or they have not read the textbooks (see marsh orchids). Then comes the problem of deciding what they are doing (see breeding ducks); or the simple factor which turns out to be incredibly complex (see water level). Methods can be a minefield – they can be critically affected by timing (see weekend effect); there are an infinite variety of methods for censusing birds (see point-time counts); and some are so intrusive they obscure more than they reveal (see cormorant predation on trout). The most unlikely-looking old data can sometimes be useful (see Perivale Wood flora); and really good-looking data can be really bad (see duck counts); at other times, data from many sources needs to be integrated to make sense of it (see palaeo-ecology and archaeology in the Lee Valley). Finally, there is an application particularly dear to my heart, and my job (see monitoring as an aid to reserve management).

10.2 FIELD MONITORING IN THEORY

10.2.1 Inadequate monitoring

I am not convinced that survey and monitoring are really different (see Chapter 1). Monitoring is usually surveying over time: a series of surveys, repeated to detect changes. If enough is known about how any survey was done, however long ago, it can be repeated and converted into 'monitoring'. Even a single survey is rarely instantaneous, and if it takes time, it is a form of monitoring. But I am convinced that a lot of it is generally inefficient and not useful.

There is a considerable amount of field recording of wildlife; much of it is systematic, and has been going on for a long time, especially on nature reserves and at observatories. There are distribution atlases, annual reports, species lists, censuses, weather records, soil temperatures, details of habitat management, and all kinds of information. But most of it tells us only that lots of people are keeping lots of records: often for no good reason, using dubious methods, and producing vast quantities of un-analysed, and often un-analysable, data. Try collating existing data from a number of different nature reserves on some species and this immediately becomes apparent.

This is often because monitoring is planned backwards, on the 'collect-now (data), think-later (of a useful question)', principle, and its components are not recognised for what they are.

Consider the following (using the classic lynx-hare population cycles as an example). Monitoring involves the systematic collection of data over time (e.g. the Hudson Bay Company's 200 years of records of pelts). The data are hard facts (e.g. numbers of lynx and snowshoe hare). From a collection of hard facts, we can derive patterns (e.g. lynx and hare numbers go up and down together, but lynx lags behind hare). Patterns indicate *what* is happening. By correlating the patterns with possible causal factors, we can estimate which are the most probable causes of the patterns – the *why* (e.g. lynx numbers follow hare's because lynx depend on hares for food). By experimental manipulations, we can then test our assumptions.

Unfortunately, data are not hard facts; pattern recognition is as much an art as a science; correlation requires suitable data on the possible causes as well as the effects; and experimental manipulations are often impractical or impossible. Thinking later may be too late to remedy deficiencies. In the lynx-hare example, the only 'hard' facts may be the number of pelts – which could be an artifact of Indian hunting strategies and unrelated to real lynx and hare numbers (e.g. see May 1980).

10.2.2 A format for good monitoring

Three questions – why, what and how, and in that order – need to be answered before any monitoring is done.

1. *Why* are you monitoring? Collecting data is not a sufficient reason in itself. You need to specify the purpose of the data, i.e. what is the question you need answering?
2. *What* data, tests and analyses will be needed to answer your question? This includes determining the type of data required, sample sizes, and appropriate statistical analyses.
3. *How* can you get such data? What methods will you need to use? This includes deciding on the sampling techniques, and any experimental manipulations; preliminary field trials will often be required to answer this question.

10.2.3 Typical formats for 'bad' monitoring

In practice, much monitoring is either descriptive-hypothetical (i.e. record-and-guess), for example:

1. collect data (e.g. bittern records at Rye Meads);
2. look for patterns (e.g. there was a bittern record once a decade before 1960, but annually after 1970);
3. think of possible causes (e.g. more regular observation since 1970 ... or more bitterns ... more habitat ... more hard winters ... etc.).

or it is *post-hoc*-correlative (i.e. see-what-fits-afterwards), for example:

1. collect data (e.g. duck counts in the Lee Valley);
2. think of a question (e.g. is the valley a hard-weather refuge?);
3. See if the data fits an answer (e.g. are duck numbers higher in the years you think were 'hard'?).

The last two approaches are extremely inefficient – lots of data collection (most of it irrelevant), and crucial information not recorded. They have just one redeeming feature: we do not yet know all the right questions to ask, and these approaches might reveal patterns which suggest some of them. Given our present ignorance of ecological systems, this is a potential benefit so important that it justifies a certain amount of 'aimless' monitoring.

It is also possible to slightly moderate the drawbacks of this collect-now, think-later approach to monitoring, by collecting data in a form and context that improves their potential, retrospective value. For example, bird data for a site may be more useful in map than table form, because maps offer more chance of relating records to other environmental variables that one might be able to deduce or go back and measure in the future.

10.3 THE METAPHYSICS OF MONITORING

10.3.1 Limitations of data

Why are data so often of minimal value? Monitoring everything is impossible. It is impossible in *theory* because we do not know enough about natural systems to know all the aspects we could record – and new techniques and approaches are being developed all the time. It is impossible in *practice* because there will never be enough resources – time, money, equipment, expertise – to record everything. Therefore, data collection is necessarily selective. This means that an assemblage of data is not objective fact: rather, it is a particular view of objective fact; and viewpoints change with time as knowledge and theory progress – which is why past data is of dubious value in answering future questions.

10.3.2 Environments

If a record is to be of any use, it must include information on its environment; and there are two crucial 'environments' to consider: the organism's environment, and the record's environment.

(a) The organism's environment

The organism has an 'environment'. Its environment is an integral part of itself. Take a robin. What is the essence of robin-ness? A specimen in a museum may be labelled 'robin', but it is not. There are differences, not just of degree but of kind, between a museum skin, a pickled whole robin, a live captive robin in a cage, and a wild robin.

A wild robin is an interacting component of the environment, a product of its past and a contributor to its future. It possesses qualities and capabilities which the other alleged 'robins' do not. Its interactive environment is an essential part of itself.

Much data collection totally ignores this point and is of the 'isolated organism' form – e.g. 'one bittern, on 15 January 1986, at Rye House Marsh'. Such a record conveys no information whatsoever on the organism's environment – e.g. why the bittern occurred (the address is inadequate). Nor does it give the record's environment.

(b) The record's environment

If you record an organism, you have engaged in an interaction – even if the organism did not notice. Such monitoring data is observer/organism interaction data, not organism data. There is a significant difference. The nature of this interaction constitutes the record's 'environment'.

The 'one bittern, on 15 January 1986, at Rye House Marsh' type of record is an almost useless – albeit common – format, because it includes no information about the probability – and therefore the significance – of that interaction. For example, if site X had 36 days on which bitterns were recorded in a year, and site Y only 12, then site X might seem three times as good for bitterns. But if site X was observed daily to get those 36 records, then bitterns were seen on only 10% of possible days; whereas if site Y was observed only monthly, bitterns were seen there on every possible (100%) occasion – indicating site X is an order of magnitude worse for bitterns.

10.3.3 Annual bird reports

Annual county bird reports bring together many observers' records. Such monitoring has been used to indicate long-term changes in bird populations (Parslow 1973). But such reports almost never give the environment of their data. The records are affected by differential site coverage, by observer bias and identification errors, by selectivity in which records are submitted, and which are accepted and published. As a result, it is impossible to draw useful conclusions from the reports alone. For example, blackbird is in the systematic list in the annual London Bird Reports in only four years from 1940 to 1960, but annually thereafter – even though we know from other sources that it was present throughout the period. Bird reports may be interesting and popular, but they have major limitations as a source of information on changes in species status.

10.3.4 Bird observatories

The situation may not be improved even when the recording coverage is known. Bird observatories collate records on a daily basis at an evening roll-call where all the observers are present; the number of observers is known, the area they have covered is known. But to compensate for observer error and bias, bird observatory wardens 'negotiate' the bird log records, so that visiting observers' records correspond more closely with the warden's expectations. Visiting observers can certainly be very unreliable; but how do we check the reliability of the warden's opinions?

10.4 MONITORING MOTIVES AND PHILOSOPHY

10.4.1 The nature of monitoring data

Monitoring provides a *particular view* of reality; but some hold that it is a direct representation of reality; others, that reality itself is a matter of opinion. These are not just philosophical irrelevancies because survey and monitoring work is being done from such different standpoints, and the standpoint affects the results.

Ecological information from conservation bodies is often a significant consideration in approving planning applications; it may all look the same, but can be anything from objective science to selective propaganda. Planning authorities often lack the expertise to perceive this distinction, and uncritically accept the results of such monitoring as objective scientific accounts – or dismiss it all as mere propaganda.

(a) Monitoring data as reality
Naive realists argue that monitoring data is a direct representation of reality. Muir (1981) considered that the scientific value of a statutory Site of Special Scientific Interest lies in the information it yields to study, and 'once this information is recorded and published whatever value remains in the objects or phenomena is of no value to science, for unless the original study was incompetently executed nothing new will be learned by preservation'.

(b) Monitoring as a political expedient
In contrast, political activists may see reality as a matter of opinion, objectivity as impossible, and therefore monitoring data as a manifestation of political ideology. On this view, monitoring data should be moulded to support the cause (the contrary view is that monitoring data should be used to determine which cause to support!).

10.4.2 Walthamstow Marshes

When Walthamstow Marshes in London were threatened by gravel extraction, the 'Save the Marshes' campaign concentrated on the nature conservation value of the site, and presented much data to support their case. But separately, a leading campaigner wrote: 'I personally don't believe that objectivity exists ... no "objective" work transcends subjective motive' (Wurzell 1986 pers. comm.); and '... scientific knowledge must be heavily spiced with sensitively calculated diplomacy... Recognizing the myth of absolute objectivity ... one always composes ones messages such that their associated vibrations and overtones are also faithful to one's intentions' (Wurzell 1980).

Given the '... wealth of authentic Marshes field data, including long bird lists' which the Save the Marshes Campaign had presented, there was some surprise at the lack of support from other conservation bodies (Wurzell 1982). But an alternative view of the 'long bird lists' is that they included contentious identifications, a third of the species were not seen on the site but only on the reservoirs next door, a further six species were just flying over, others were confined to the neighbouring river, and some were 'races' so that some species were counted more than once (e.g. Harvey 1982).

10.5 OBSERVATION PROBLEMS

10.5.1 Tropical rain forest birds

Trying to census birds in an unexplored area of tropical rain forest in the 'Lost World' Guyana Highlands in 1979 posed two particular problems: seeing the birds, and identifying them.

Dubious identifications render all the work suspect. Field identification was essential as we wished to observe behaviour rather than collect specimens. The British Museum (Natural History) kindly gave us access to their skin collections; and fortunately a very good field guide to neighbouring Venezuela (de Schauensee and Phelps 1978) had been published a few months earlier.

Once in the rain forest, the problem was seeing the birds. The majority were up in the forest canopy. The canopy was 40 m above the ground. The alluvial forest in the study area was so dense we could only see a small bit of the canopy from any point. Any birds up there were silhouetted.

We had to get up into the canopy. We used a catapult and lead weight to fire a nylon fishing line over a branch, then hauled up successively stronger ropes until we could tie off and climb the rope using ascenders. By repeating this process we eventually reached the canopy. Once there, we had to be very careful about touching any lianes: ants were abundant in the trees, used lianes as highways, and often built their nests around them: they took very unkindly to being disturbed. For the same reason, we had to leave the rope end hanging free of the ground so that the ants did not adopt it as a route.

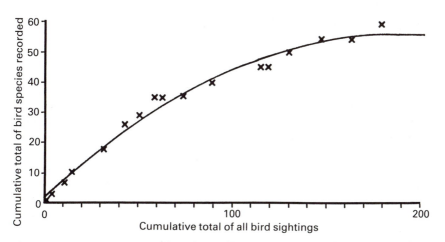

Figure 10.1 Tree-top census-data from alluvial tropical rain forest 500 m above sea-level at the foot of Mt. Eborupu, Guyana, in 1979.

Apart from vertigo and ants, we still had other problems. The canopy was very thick. Most of the forest birds moved around in flocks. Flocks might consist of 100 individuals of perhaps 30 species. One might wait an hour for such a flock to appear, and it would be around for ten minutes or less. Even then, most birds would be obscured from view most of the time.

We tried to gain a few extra seconds of visibility by making an opening in the canopy. The first (and only) tree we tried felling still stood there after we had chopped clear through it, so well-supported was it by its neighbours and lianes. By ramming it off its stump, we succeeded in making the cut base descend to the ground; and by continuously cutting off successive sections from the bottom of the trunk, we finally (after two days!) got the tree lowered out of the canopy.

We then monitored birds from our observation branch until after several days the number of new species we were recording levelled off, suggesting we had recorded most of the species in the area and so could move on to a new site (Figure 10.1).

10.6 IDENTIFICATION PROBLEMS

10.6.1 Marsh orchid swarms

Choosing the units to record is not always straightforward. Birds are discrete organisms, and most are usually distinct species. Plants are less accommodating. Vegetative reproduction means that the concept of an individual plant is not always useful; and species identification can be confounded by much high frequencies of hybridisation.

We needed to monitor population changes in 'hybrid swarms' of *Dactylorhiza* orchids on fly-ash at Turnford, because the orchids were of conservation value and increasing interest to visitors, but a catastrophic population decline had been reported from another Lee Valley site (see also Chapter 4).

The first problem with monitoring numbers was classifying individual plants. Many looked nothing like standard *D. incarnata* (L.) Soo, *D. majalis* (Reichenb.) P.F. Hunt and Summerhayes subsp. *praetermissa* (Druce) D. Moresby Moore and Soo, or *D. fuchsii* (Druce) Soo. With these marsh orchid swarms, prejudging the taxa along conventional lines and calling anything inconvenient a 'hybrid', seemed inadequate. It seemed more appropriate to study character variation across the whole population and use that to define taxa. Accordingly, flower characters such as colour, sepal position, spur shape, and labellum markings, length, folding and lobing were recorded from a random sample of orchids, and the results subjected to cluster analysis.

This gave us a classification of distinct forms whose populations we could

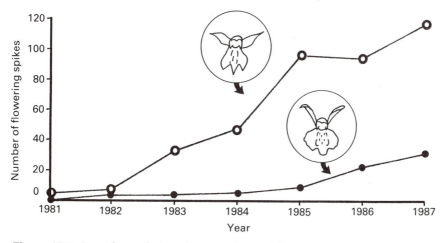

Figure 10.2 Annual population changes of two different morphs (M30: hollow circles, M28 solid circles) of the *Dactylorhiza fuchsii* complex at Cheshunt, Hertfordshire, 1981 to 1987.

monitor and habitat preferences we could plot. There were major divisions in line with conventional species concepts – a *cf D. fuchsii* cluster, a *cf D. majalis praetermissa* cluster, and two *cf D.incarnata* clusters – but also many distinct sub-clusters which showed very different preferences and trends.

For example, within the *cf D.majalis praetermissa* division, one morph showed a rapid population increase and a preference for open glades, whereas another had a preference for shade and showed a slow increase in population followed by a gradual decline; and two morphs in the *cf D.fuchsii* division with shade preferences showed very different rates of population increase (Figure 10.2).

10.7 SELECTING THE CRITERIA

10.7.1 Duck breeding success

Often there are a number of different criteria that could be monitored to answer a particular question. Determining the best breeding site for tufted ducks *Aythya fuligula* in the Lee Valley could be done by monitoring numbers of breeding pairs, or nests, or young. But different criteria can indicate very different answers. G. White's (*in prep.*) careful work (Table 10.1) in the Lee Valley demonstrated the importance of selecting the right criteria, and the value of monitoring to evaluate those criteria.

His results showed that Walthamstow had most pairs, over half-as-many again as the next-best site (Amwell); Amwell had most nests, about a third

Table 10.1 Breeding tufted duck *Aythya fuligula* in the Lee Valley, 1986

	Pairs*	Nesting	Broods	Fledged
Rye Meads	68–77	53	61	221
Cheshunt	30–48	36	29	30
Amwell	67–84	70	62	58
Walthamstow	105–127	26	15	12

* Pairs = number of pairs on the water immediately prior to the onset of incubation in late May; nesting = number of females disappearing off the water in late May; broods = number of separate broods on the water; fledged = total number of young seen fledged (data courtesy of Graham White).

more than the next-best site (Rye meads); Rye Meads and Amwell equally held most broods, both with over twice as many as the next-best site (Cheshunt); but Rye Meads had far and away the most ducklings, nearly four times the next-best site (Amwell); and in terms of number of young per pair, Cheshunt, which ranked third or fourth on all the other criteria, came second-best.

Each criterion produced a different ranking of sites. Only by monitoring through the season could the full picture be determined; even then, that season may not be representative of site productivity in other years. Even the number of young produced, or the number of those reaching maturity, may not be relevant. It may be that a few young from one site could end up contributing a lot more offspring to future generations than a lot of young from another site. The contribution of individuals to the gene pool is highly skewed, and site averages may be very poor indicators.

10.8 TIMING

10.8.1 The weekend effect

One day is not necessarily as good as another. Most rare birds appear in Britain at the weekend – a fact that tells more about bird-watchers than birds. Much monitoring takes place at weekends because it is the only free time many people have. Duck counts tend to be on Sundays. Sundays are not a typical day of the week: there is much more use of waters for water-sports on Sundays. The national duck counts are therefore likely to be particularly atypical representations of normal duck distributions.

This argument persuaded the Nature Conservancy Council to fund a study on Walthamstow Reservoirs. However, a comparison of weekday and weekend counts on all 11 basins throughout the 1987–8 winter failed to find any significant differences.

On the other hand, unfunded observations on the two basins of the King George V reservoir did show a significant weekend effect, because water sports there in the 1985–6 and 1986–7 winters were almost exclusively (92%) confined to weekends.

Separate duck counts and water sports activities records were kept for each basin. Water-sports on one basin resulted in significant declines in tufted duck and shoveler numbers on that basin that day; but numbers on the site as a whole showed no significant change as long as the other basin remained undisturbed. However, if water-sports occurred on both basins simultaneously, then both tufted duck and shoveler showed highly significant reductions in numbers over the whole site.

These results depended on monitoring relevant factors at the same time as the species being studied, and recording them with sufficient precision – in this case, per basin. A comparison of weekend and weekday counts would have suggested weaker associations, because water-sports occurred on only 79% of weekends, and on only 43% were both basins disturbed. Lumping duck counts and water sports records for both basins would have completely concealed the impact on duck numbers because water sports only affected site totals on the days that both basins were used.

10.9 SAMPLES AND REALITY

10.9.1 Bird censuses

There has been a great deal of practical and theoretical work on bird censuses and there are a number of collections of useful papers on the subject (e.g. Oelkel 1980; Taylor *et al.* 1983; Ralph and Scott 1981).

Determining real densities from census samples is a major problem. Some techniques avoid the problem by merely providing indices of relative change. If an observer records 100 blackbirds in a woodland, this may be a fraction of the real population, or a gross overestimate. Nevertheless, if the same observer using the same technique later records only 50 blackbirds in the same wood, it is a reasonable presumption that blackbird numbers there have declined.

Such conclusions are particularly valid if similar results are produced independently by other observers on other sites. This is the principle of the Common Birds Census (CBC) mapping technique for breeding populations. The relationships of CBC results to real densities are variable and obscure, but it is an effective technique for indicating relative trends (see Chapter 7).

10.9.2 Transect and point count techniques

In transect censuses, birds are recorded in a strip along a fixed route representative of the habitat being sampled; and the sighting angle and distance of each bird are noted. A minimum requirement of 40 individuals of each species has been suggested (Burnham *et al.* 1981).

The point count technique (Blondel *et al.* 1970), and its myriad variants, involves recording all the birds seen in a small area within a certain time. For example, at various points or times along a route the observer stops and records every bird, either within a fixed distance, or estimating the distance of each sighting. A minimum of 20 stations with 100 m between each station has been suggested.

Density estimates from these techniques can be confounded by activity levels. One bird could cross the transect route many times and be recorded many times. Point counts in particular can be more indicative of activity than density.

The other problem is detectability: all techniques require some estimate of the detectability of each species in the census area; but uncorrected census results are often presented – as if the observer has actually detected all the birds present! This is self-delusion.

10.9.3 Alternative approaches

Most census techniques involve sampling both the habitat (you select particular census areas), and the birds (you detect and record as many of the birds as possible in your sample plots). You then have a double problem of determining the relationship of your samples to reality.

Alternatively, you could pick a block of habitat so small that you could record the total population of birds in it. The only problem then is determining the relationship of your sample block to the rest of the habitat. But at least your bird numbers are real, not estimates.

In open habitats such as fields of short grass, such total censuses become a possibility. In closed habitats such as reed-beds and scrub the sample plots may have to be very small indeed to record everything. The smaller the plot, the less representative it is likely to be. Inside *Phragmites* reed-beds in northern Greece in 1975 we had maximum visibility of 0.3 m – at which distance any contacts were a severe shock to both bird and observer. Such extreme conditions require a different approach.

10.9.4 Point-time area (PTA) counts

The point-time area count aims to record all the birds present in a plot at a point in time. Because each census detects everything at a point – instant – in time, it gives a real density (population per unit area), not an estimate. Either all the birds present in the plot must be observable at the same instant, or the site must be watched for long enough after the point-in-time to detect all the

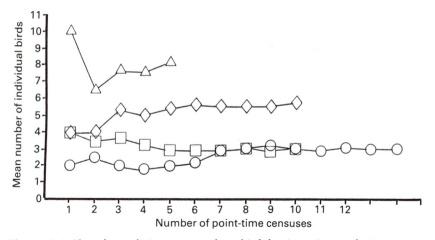

Figure 10.3 Plot of cumulative averages from bird density estimates during autumn passage, using series of point-time counts from the same station in fen-carr habitat on Rye House Marsh RSPB reserve, Hertfordshire. The plots show progressive reduction in variance with increasing numbers of censuses. Triangles: 1900 hrs on 9 August 1982; diamonds: 0900 hrs on 15 August 1982; squares: 1500 hrs on 15 August 1982; circles: 1100 hrs on 14 August 1982.

birds that were in the plot, and all the birds entering after the census point must be detected, so they can be omitted from the results.

The whole census plot must be in view at the same instant, so plots might need to measure as little as 25 × 25 m. Point-counts should be repeated in the same spot at random time intervals until the running-average density estimate stabilises (see Figure 10.3). The temptation to cheat and census only when lots of birds are in view must be avoided; if done properly, many censuses should record no birds.

If the vegetation is too thick to see all birds present, the point-time density has to be estimated from an extended observation period – the optimum time can be estimated from the rate of decline of new contacts. Identifications should be as fine as possible: e.g. two blackcap sightings might be the same bird, but if one was male and the other female it is two birds.

The population at the census point includes any bird seen or heard in or leaving the plot, unless it entered after the census point; the precise sequence of each observation, entry and exit is critical for estimating the minimum population present. Such data also indicates the 'flow rate' of birds through the habitat. Detecting movements in and out of the sample plot is essential but difficult. Seeing all round a plot is difficult – elevation helps, assistants are invaluable. In Greece, goat tracks provided sight-lines – wide gaps may produce an 'island' effect.

In Greece in 1975 such an approach allowed us to demonstrate changes in habitat occupancy by temperate-zone warblers on passage which were closely related to the presence of competing species – as Mediterranean summer residents differentially vacated each habitat, northern passage migrant densities selectively increased in them.

At the RSPB's Rye House Marsh in 1981 we were able to calculate that through August there were mean densities of 53.8 passerines per hectare in fen-carr habitat on the Rye House Marsh reserve, which meant 123 birds at any one time in our 2.28 ha; but flow-rate measurements showed this comprised an average of 2298 birds per day foraging through the reserve.

10.10 SIMPLE FACTORS WITH COMPLEX CONSEQUENCES

10.10.1 Water levels

Water level is easily monitored. It is a simple linear variable. But it has profoundly non-linear and extremely complex ecological consequences. Monitoring water levels can turn out to be the tip of the iceberg as far as necessary data are concerned.

At Rye House Marsh water comes into the various compartments from a feeder ditch on one side of the site, and drains out into the river on the other. The fall is less than 1 m. Where possible, horizontal 150 mm diameter drainpipes were installed from the lowest point in each compartment into the feeder ditch, where a right-angle bend allowed them to be extended above the feeder ditch surface. The level in each compartment could then be determined as the difference between the water in the pipe and the ditch, and the pipes could be used to let water in.

For baseline measurements we used metre rules of plastic so the water would not swell or rot them, and with the intervals impressed so they would not rub off. These were all placed along a screened route so we could read them without disturbing birds on the reserve. The 1 m mark was at the top and at exactly the same height – using the water itself as a level – in each location, to assist correlating records across the site. Site datum was chosen so that it was below the minimum height of the river into which we drained.

10.10.2 Complex effects

Similar rises in water level required very different volumes of water, because the cross-ditches widened upwards; and once the top of the ditch was reached, the water spread across the scrapes and a very much greater volume was required to raise the level.

Exposed mud area remained similar while water was rising in the ditch, but

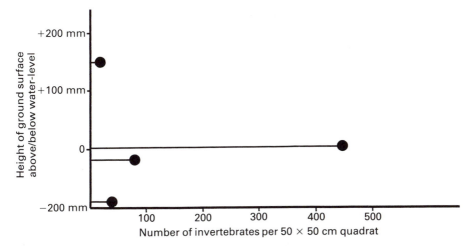

Figure 10.4 Numbers of invertebrates in equivalent samples of mud in relation to water-table height, scrape habitat, Rye House Marsh RSPB reserve, 15 August 1989.

decreased very quickly as water overflowed across the scrape, stabilising briefly while a few hummocks remained above the surface.

Invertebrate numbers varied with the water level. With low water levels, the surface was dry and numbers were low (Figure 10.4). As levels rose, numbers increased dramatically, reaching a peak when the water was within 20 mm of the surface. As flooding proceeded, invertebrate numbers showed an equally dramatic decline.

Birds feeding were affected by numbers of invertebrates, water depth and hardness of the mud. Snipe, with their long, thin bills could only probe in a very soft substrate (Figure 10.5). Mud hardness changed little once under water, but hardened rapidly as water levels fell below the surface. Birds such as snipe feed preferentially on the water margins; margins moved but margin length remained relatively constant, except briefly when the scrapes first started to flood, when there was a massive increase as many pools formed.

Most wading birds only feed in water depths less than their bill or leg-length – whichever is the shortest – so that their plumage does not become soaked. At greater depths, dabbling ducks feed while floating. Neck length determines preferred feeding depth; up-ending is energetically more expensive but increases feeding depths to approximately half the body, plus the neck-length.

Changing water levels also result in activity-switching. Snipe are absent from the scrapes at Rye House when they are dry; they feed there during shallow floods; and use the few exposed hummocks as roosts when the water is high.

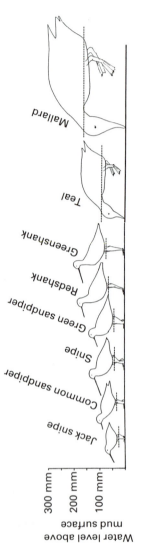

Mallard
Teal
Greenshank
Redshank
Green sandpiper
Snipe
Common sandpiper
Jack snipe

300 mm
200 mm
100 mm

Water level above
mud surface

Figure 10.5 Maximum usable water depths by various species of ducks and waders on freshwater scrapes at Rye House Marsh RSPB reserve, Hertfordshire.

Temporal patterns introduce further complications. Water levels may vary hourly, or daily, or seasonally. Time-span is one aspect, timing another: winter floods have different effects (e.g. on flora, invertebrates etc.) to summer floods. Even constant water levels have a time dimension – the older a water, the bigger fish may grow; or the more duck will find it and traditionally return every year, so that numbers increase over time precisely because the environment has not changed.

10.11 TO INTERFERE – OR NOT?

10.11.1 A predator-prey relationship: cormorants and trout

Studying predation by predating the predator poses a problem: it interferes catastrophically with the predator-prey interaction system.

The effect of cormorants on fish stocks is often assessed by shooting the cormorants and analysing their stomach contents (e.g. Mills 1965); but in 1982 we got an opportunity to use an observational approach, when Thames Water alleged they had a cormorant problem on their Walthamstow trout fishery because theoretical trout stock density (what they put in, less what anglers took out) could only account for 16% of the variance in anglers' catch-rates, and cormorants could be seen catching trout (Figure 10.6).

We knew how many trout were put in, of what size, and when, because the fisheries staff kept a stocking record. We knew how many the anglers took

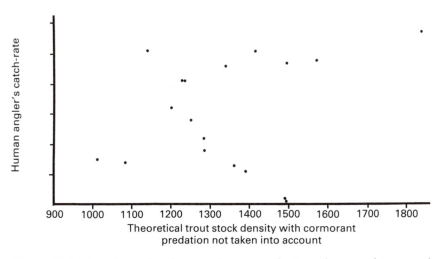

Figure 10.6 The relationship between human anglers' catch-rates of trout and numbers of trout released into Walthamstow Reservoirs in 1981, without allowing for the effects of cormorant predation.

out, because they had to report their catches and there were spot-checks. But we did not know how many the cormorants took out, or how many trout were lost for other reasons; and there was no way of censusing the trout once they were in the reservoir.

The number of cormorants fishing, and the number of trout each ate, were both potentially variables. Both were difficult to determine as the reservoir was 17.4 hectares; there was a lot of coming and going; birds were often underwater and so invisible; and fish were swallowed within seconds of being brought to the surface – and getting there in time for the dawn 'feeding frenzy' meant coping with the London rush-hour traffic, a variable we sometimes misjudged.

After some trial and error, numbers on the fishery were estimated from the highest surface counts (even by chance, there will be times when most will be on the surface at the same time); and time on the fishery from timed records of all birds flying in and out.

Fishing effort and success were determined by recording within a fixed field of view all dives per minute, dive lengths, and birds surfacing with a fish and swallowing it. (We did not use surfaces-without-a-fish data because in a flock, it was much easier to be sure that a cormorant had just disappeared from view, than it was to be sure that one had just appeared – unless it had a fish. Either that, or our data show a lot more cormorants dived than surfaced ...)

These observations allowed us to produce a far more powerful predictive model than just cormorant counts and stomach analyses would have allowed (Figure 10.7). Our data showed that the numbers of cormorants on the fishery varied, first as more and more birds discovered that trout were being stocked, then in a density-dependent fashion related to trout numbers (fewer cormorants bothered to fish there at low stocking densities); and not all cormorants on the fishery predated trout (some specialised in sticklebacks and had a quite different feeding action). Also, cormorants ate one 370 g trout per day on average, but lack of experience meant that it took several days of practice for most birds to achieve that.

The model allowed us to estimate daily trout stock densities in the reservoir. There was no way of counting the trout left in the reservoir, but anglers' daily catch-rates should have shown a close relationship to stock density: and our model was able to explain 78% of the total variance.

We had hoped to find that cormorants had no significant effect on anglers' catches. Unfortunately, our research confirmed the Water Authority's prejudice, even though cormorants took less fish than was thought. Our other studies indicated that culling would be ineffective because of the transient nature and large number of individual cormorants passing through. In the end we concluded a 'time-share' arrangement. The winter trout fishery ceased, and cormorants were allowed to use the reservoir till the trout fishery

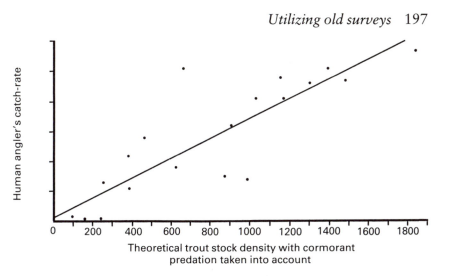

Figure 10.7 The relationship between human anglers' catch-rates of trout and numbers of trout released into Walthamstow Reservoirs in 1982, allowing for the effects of cormorant predation using our model.

reopened in March. Crop-scarers below the roost persuaded the cormorants to accept this arrangement and leave for the coastal breeding sites a few weeks earlier than they would have liked.

10.12 UTILIZING OLD SURVEYS

10.12.1 Alchemy: converting old surveys into useful monitoring

Perivale Wood in north-west London is an isolated island of ancient woodland surrounded by London suburbs. It has been the subject of three floristic surveys this century – 1913 (Shenstone 1913), 1961 (Groves 1962), and 1973 (Roberts and Edwards 1974). Potentially, these could provide valuable information on changes over 60 years, but most of them only published their results, rather than their methods – thus they are difficult to compare. Any differences in the results may be due to differences in the thoroughness of the surveys.

The total numbers of species recorded was very similar in each survey (74 in 1913, 73 in 1973). This might suggest some stability in the woodland flora over the last 60 years; but it almost certainly reflects higher species-detection rates in later surveys, which conceal significant changes.

One way of using such data is to examine species persistence over time: taking the first survey as a baseline, the later surveys were checked to see how many of the 1913 species were still present in 1961 and 1973. These results

Table 10.2 The number of flowering plant species recorded in 1913 at Perivale Wood, north-west London, which were still recorded present in the surveys of 1961 and 1973

Survey year	1913	1961	1973
Ecological category			
Trees and shrubs	11	10	11
Climbers and creepers	4	4	4
Shade herbs	21	13	10
Meadow species	14	0	2
Waste-place species	16	5	3
Total	66	32	30

(Table 10.2) show all tree and shrub species, and climbers and creepers, were still present after 60 years; but shade herbs show a reduction of some 50%; and meadow and waste-place species a reduction of 83%. The more light-demanding and less-competitive the group, the more species have been lost during the period – precisely the effect expected as coppice management lapsed after the First World War and the canopy closed.

10.13 A PROBLEM WITH NUMBERS

10.13.1 Wintering wildfowl

Wintering wildfowl are well-monitored in the UK. Ducks are relatively easy to count on open waters if you have binoculars or a telescope and the ability to identify the different species. There have been systematic counts for a long time. We have lots of figures. We can do lots of statistics. I have a typical report (Anderson and Edwards 1985) that lists monthly average counts of wildfowl on waters in the Lee Valley, including for example the fact that Amwell Gravel Pits hold an average of 0.888889 shoveler in January .... six decimal places of a shoveler duck.

Such data provide a strong temptation to think we have some quite precise data on ducks – and that we know something about them. On a national scale, perhaps we do. On a local scale, at which the counts are done, this is a very dubious assumption.

10.13.2 National Wildfowl Counts

Since 1947, there have been synchronised monthly counts of wildfowl on selected waters in the UK. The purpose of this scheme was to determine the status of wildfowl and any long-term trends (Atkinson-Willes 1963). At a national level, such trends are likely to be 'real'. Numbers on single waters fluctuate, but if many waters are counted such fluctuations cancel out and consistent trends emerge. Observers' counts vary in accuracy, but there seems to be no bias so again errors cancel out (Atkinson-Willes 1963).

One use of this data is to assess the importance of sites – those that hold more than 1% of the national population of a species are conventionally considered of national significance. The logical flaw is that the counts that comprise the national totals are not reliable indicators of the populations on a single site, because the error-cancelling when counts are accumulated at a national level no longer applies.

Several ways round this have been proposed (Atkinson-Willes 1963). Average counts are poor indicators, due to local fluctuations, atypical counts resulting from adverse weather or disturbance, and missing data. Alternatives are the regular population, defined as the average of the three highest counts for each season for which adequate records (at least five of the seven monthly counts) exist; the maximum population, defined as the largest single count in the study period; or the average peak, defined as the average of the highest single count of each season under review.

Statistically, these methods average out some of the errors. What they do not do is address the problem of whether the data are relevant to the question. These problems were highlighted during assessments of whether the King George V reservoir in the Lee Valley merited SSSI status (Roberts and White 1985).

Average maximum goosander counts on the King George in the 1970s and 1980s numbered 20–30 birds – below the 50 required for national significance, and an apparently clear-cut, quantitative conclusion. Such attempts to answer questions with irrelevant data are commonplace.

In reality, goosander in the valley used two reservoirs – the King George and the nearby Girling – and moved freely between them; birds on both were effectively part of the same population, which averaged 50.3 birds – the level required for national significance. Even the two reservoirs were not sufficient, because the critical goosander feeding sites were the adjacent flood-relief channels, which the birds could only use at dawn without being disturbed, and where they were never detected by conventional duck counts.

The availability of detailed duck counts has tended to obscure the fact that we know almost nothing about wintering wildfowl. Conservationists have usually evaluated water bodies in isolation; ducks are more intelligent and make strategic use of different sites for different purposes. Single waters only have value if they are part of a complete system; and a single water can rarely supply all requirements.

In winter, for example, ducks are likely to require a feeding site (possibly more than one if food stocks are to last all winter), a roost site, a temporary refuge in the event of disturbance, and sometimes a hard-weather refuge.

We examined this problem in the Lee Valley (Roberts and White 1988) by counting at different times of day and night throughout the week, recording what the ducks were doing (feeding, roosting etc.), and any other factors potentially affecting the counts (ice, boating etc.).

This showed that conventional wildfowl counts were detecting only fragmentary and diverse aspects of wildfowl strategic use patterns. Feeding areas of several species were not detected at all: all the mallard concentrations were roosts; of three teal concentrations, two were roosts, the other (underestimated) a disturbance refuge; the goosander concentrations were also roost/refuges.

Several different concentrations identified by conventional counts also turned out to refer to the same birds: e.g. the Broxbourne and Rye Meads shoveler concentrations were the roosting and feeding sites respectively of the same birds. Goosander and goldeneye seen on the King George and William Girling reservoirs were also the same birds, using the former site as a roost and the latter as a disturbance refuge.

10.14 INTEGRATED MONITORING

10.14.1 Palaeo-ecology, archaeology and aerial photography

The concept of monitoring is equally applicable to 'fossilised' systems which are relatively unchanging over time, except for their availability to study. In 1987 a new gravel pit at Rye Meads in the Lee Valley floodplain provided an opportunity to investigate on the same site the late- and post-glacial stratigraphy, an extensive mesolithic occupation site, and the present vegetation patterns. Each required a different approach, but each threw light on the others, as well as elucidating concepts of relevance to present day conservation management, such as naturalness and natural change.

The present vegetation patterns reflected the buried topography, and were invaluable for relating the stratigraphy exposed in different parts of the site; this made it possible to interpret the past site topography, which proved crucial to understanding the differential distribution of mesolithic material and the ways they used different parts of the site (Figure 10.8).

Monitoring was a problem. Ninety per cent of the site was destroyed within nine months. Many exposures only lasted overnight and were not accessible in daylight; were only visible from the deep, water-filled pit; and some were extremely dangerous to approach because of mounds of overburden stacked above them. An extremely wet summer exacerbated these problems, increasing the dangers of collapse, and often flooding the sections despite continuous pumping. The site had to be monitored frequently, and with great attention to safety. Each aspect of the study required a different approach.

10.14.2 Present vegetation

The field before gravel extraction was flat, damp, and with a uniform surface substrate of thick flood clay. However, aerial photographs showed distinct vegetation patterns; and on excavation, these were found to precisely reflect the buried late-glacial landscape of torrent-gravel ridges and hollows, even though the original vertical range of over 4000 mm some 25 000 years ago had been reduced by an order of magnitude by subsequent infilling, and the whole site then capped by flood-clay.

Patterns were drawn from the photograph, then checked in the field. Key indicator species were lesser pond sedge *Carex acutiformis* which mapped the old buried river channel; and soft rush *Juncus effusus* which dominated a shallow lake site and hollows. Tufted hair grass *Deschampsia cespitosa* characterised the gravel slopes, and the ridge-tops supported grasses and occasional woody plants.

This close mapping of present vegetation to buried landform was probably due to the sensitivity of marshland plant species to small differences in water levels and thus surface height. Even so, the patterns were only recognisable on aerial photographs taken at certain times of the year.

10.14.3 Stratigraphy

Determining the stratigraphy necessitated recording exact altitude and location of each section; description and measurement of all strata; any flora, fauna or artifacts; and evidence of processes (e.g. current-sorting, erosion cliffs, cryoturbation etc.); plus taking samples for later pollen analysis etc. Cleaning all sections was essential, as smearing by the draglines often gave a very misleading impression.

There was a great temptation to 'join the dots' – to assume a buried channel cross-section in one side of the pit joined up with an identical feature on the other. But the plan view provided by the aerial photographs suggested quite different patterns, which turned out to make much more sense, revealing a buried river channel, lake, ridges, closed hollows and other features, which also explained some of the patchy distributions of certain strata.

The distribution of many strata was so patchy and uneven that 76 sections were recorded before the exact relationships could be established; but eventually a full sequence of 13 strata was constructed, from late-glacial torrent gravels, an Allerod interstadial lake, Late Dryas ice-raft gravels and aeolian deposits, and then post-glacial soil development, mesolithic occupation, erosion, flooding and peat formation, and finally Iron Age flood-clay deposition, probably due to massive deforestation of the watershed.

(a)

(c)

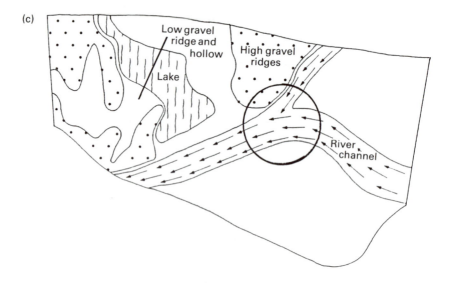

(b)

(d)

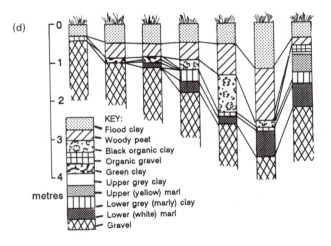

KEY:
Flood clay
Woody peat
Black organic clay
Organic gravel
Green clay
Upper grey clay
Upper (yellow) marl
Lower grey (marly) clay
Lower (white) marl
Gravel

metres

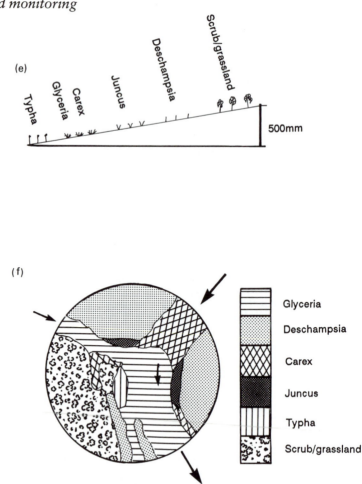

(e)

Typha
Glyceria
Carex
Juncus
Deschampsia
Scrub/grassland

500mm

(f)

Glyceria
Deschampsia
Carex
Juncus
Typha
Scrub/grassland

Figure 10.8 Relationships between stratigraphical data, ancient buried landscapes, aerial photography and contemporary vegetation cover on a flat field on the floodplain of the River Lee at Rye Meads, Hertfordshire (the site of a major mesolithic occupation).

(a) Aerial photograph.
(b) Contrast-enhanced bit-map from aerial photo.
(c) Map of main buried mesolithic landscape features, showing location of Figure 10.8(d) (line) and Figure 10.8(f) (circle).
(d) Borehole stratigraphy across the buried lake.
(e) Vegetation changes with ground-level.
(f) Present-day vegetation around the junction of the buried river channels.

10.14.4 Archaeology

The conventional approach – which the conventional archaeologists took – was intensive, excavating a few square metres in great detail before running out of time. Mesolithic material was present over some four hectares so one trench of a few square metres was hardly likely to be representative.

Instead, the gravel pit faces were treated as transects for sampling purposes. The data needed to be suitable for statistical analysis to see if differences were probably real or sampling error. Therefore samples were collected on a stratified random basis according to the different types of contemporary site topography indicated by the stratigraphic and modern vegetation studies (e.g. a riverside gravel ridge, clay hollows around a lake shore, etc.). Exact sample areas and volumes were recorded; location, depth and stratigraphic position were plotted; and all burnt flints, worked flints, bone, wood and stones other than flint were recorded.

Analysis showed a highly structured occupation. Open hearths were clustered beside the river, and had statistically different artifact assemblages, suggesting different functions (e.g. skin preparation). Enclosed hearths – suggestive of dwellings – were concentrated in shallow hollows away from the river. Other significant differences suggested most primary production of flint implements took place in the clay hollows, well away from the hearths in regular use.

10.15 BIRDS AND RESERVE MANAGEMENT

Conservationists managing reserves can feel a particular need to interfere in order to justify their employment. The effects of such habitat-molesting should be clear, because reserves get more monitoring than most sites. Monitoring invariably shows that management is 'a good thing' – but only because most monitoring is unconsciously rigged to allow no other answer.

Inadequate baseline surveys and lack of unmanaged control plots obscure any damage to the wildlife interest; monitoring is biased towards the post-management phase, so more recording inevitably detects more wildlife afterwards; predictions are vague, allowing scope for nimble footwork after the event (predictions that come true can be claimed as evidence of appropriate management, but those that do not can always be blamed on off-site factors, such as the weather!); and there is rarely any coincidence between management areas and wildlife recording areas, so it is even more difficult to relate the management to any changes in species status (knowing that nightingales have increased from five to 25 pairs after 10% of the reserve was coppiced, is not nearly as useful as also knowing whether your 20 extra pairs are actually in your coppice area).

10.15.1 Reasons for monitoring

Reserve management is expensive in terms of time and/or money. At the Rye House Marsh reserve of the Royal Society for the Protection of Birds we wanted:

1. to monitor any changes in the bird population;
2. to know how effective our management was (how much of what we saw was due to what we did ...);
3. to know how much it cost, so that we knew what to budget, and whether it was worth it;
4. to find the most cost-effective ways of achieving some of our targets (e.g. maintaining the typical bird community of willow carr, or increasing wintering snipe numbers).

We therefore set about monitoring with the aim of producing a predictive model of costs and effects on each bird species (and plants), so that we could simulate options on the computer rather than have to rely on trial-and-error in the field.

10.15.2 The site and its functions

The RSPB's Rye House Marsh reserve – all 5 ha of it – was set up primarily as an educational site for schools and colleges; it also has a major public visiting function (it is one of the 20 most visited RSPB reserves); and now has a wildlife resource conservation function as part of the larger Rye Meads statutory Site of Special Scientific Interest.

When the site – derelict flood-meadow – became a reserve in 1973 it had become a monoculture of *Glyceria maxima* due to high nutrient levels in water from the neighbouring sewage works. It was not typical of any natural habitat, and had negligible intrinsic wildlife interest.

From an educational point of view, for our taught courses, we required examples of a variety of wetland habitats to contrast and compare; with typical bird communities we could use to illustrate ecological principles. For public visiting, generally informal and unescorted, we also wanted to maximise the wildlife spectacle – i.e. opportunities to see particular species, and as many species as possible. We also wished to maintain or enhance the wildlife resource value of the site – an essential pre-requisite if our education and visiting objectives (nature conservation) were to remain credible.

We therefore created examples of the various stages of natural wetland succession: bare mud and gravel such as would have been exposed by natural river action in the past, by scraping off the surface of some areas; *Phragmites* reed-swamp and mixed fen, by reducing nutrient levels in the water by filtering through *Glyceria* beds; and fen carr, by planting willows and alders.

We also resuscitated part of the historical, agricultural landscape of flood-meadow and drainage ditches, by clearing out ditches and removing the *Glyceria* mat, to expose the old meadow soil and allow the buried seed-bank to regenerate. Finally, we created maximum-diversity areas with bits of all those habitats, and 'spectacles' such as tern nesting rafts, to enhance visitors' chances of seeing wildlife. Each of these units has specific running costs in terms of its maintenance requirements.

10.15.3 Bird recording

The first requirement was to create habitat adequate to attract typical species; then to maximise the amount of time those species spent on site; and finally to maximise their populations. Species lists alone were too crude to do more than suggest the presence of suitable habitat. As an index of species regularity, we chose species presence/absence per day. As an index of species population size, we used the maximum count per day.

Species presence/absence is a more robust measure than maximum number, but unable to detect significant population changes in regular species – one snipe or 1000 both count just as 'present'. Maximum counts overcome this drawback but are more easily distorted by differences in observer coverage or atypical events; and although they are useful for comparing changes within species, they do not necessarily allow comparisons between species – e.g. all the teal on the scrapes, but only a few of the warblers in scrub, tended to be visible at one time.

A small study area near the entrance to the reserve was selected to allow good coverage: it comprised 1.5 ha, with no visitor access but was observable for two hides.

Observer records will be influenced by factors such as time of day, area visited, expertise, and method. With one observer, any bias will be consistent and any trends in the data should be real. However, the chances of the same observer being able to do all the recording over several years are small. With a few observers, consistency is reduced and there is a significant danger that individual biases will not cancel out. If a single observer is impractical, then there should be as many observers as possible. We therefore set up a scheme whereby all visitors to the reserve at weekends were encouraged to contribute records – of the maximum numbers of each species they had seen that day in the study area – to our bird 'scoreboard'.

All the species ever seen on the reserve were listed on the scoreboard – otherwise 'uninteresting' species were rarely reported. Only records of birds in the study area were accepted – e.g. just flying over did not count as 'in' (except for things like kestrels hovering). And *maximum* count meant *minimum* number – i.e. five blackbirds seen together = five blackbirds, but five sightings of one blackbird = one blackbird, unless they were clearly different individuals, e.g. males and females. The study area could not be entered, but could be observed from hides.

10.15.4 Environmental data and experimental manipulation

The bird data could indicate what was happening, but not why. We needed simultaneous recording of on-site environmental data to indicate how much of what was happening might be ascribed to our management. The study area was established so that we could hold observability constant while varying environmental factors.

Relevant on-site factors were considered to be the area of each habitat type (woody scrub, tall reed-fen, open mud and short turf, disturbed ground); configuration; site conditions (length of recently-cleared ditch-edges, water-levels, water-flow, ice); visitor pressure (disturbance); and species interactions.

Habitat configuration – the relative positioning and shape of blocks of different habitats – is likely to be important because many birds use more than one habitat, or the ecotones between them. But there are practical difficulties with altering configurations, especially with habitats which take several years to develop (e.g. scrub), so habitats were laid out in the study area in a series of concentric rings from early successional to late, and areas were varied by altering the width of the rings (e.g. expanding the scrapes into the surrounding reed, or cutting back the edge of the scrub).

The other major group of factors affecting birds on the reserve is likely to be interactions between species. Potentially significant interactions are parasitic, such as cuckoo on reed warbler; aggressive, such as coot on teal in summer; predatory, such as great grey shrike on bearded tit in winter; and indirect, such as the effect of goose-grazing on vegetation structure, which then affects the species and numbers of birds using it.

10.15.5 Analysis and results

By varying the environmental factors over the years, recording their management costs, and simultaneously monitoring the bird populations, we were able to use multiple linear regression as a first-stage approximation to estimate how much of the changes in the bird populations could be ascribed to our management. And because we also recorded management time, we could determine the manpower costs of each option – and also detect any changes over the years (e.g. hay-cutting and flood-meadow was twice as costly the first year – because of the litter build-up etc. – as it was in later years). Ten years' data gave us 120 monthly samples of % days present per month and maximum counts for each species, or 30 samples if they were grouped in three-monthly seasons.

On-site factors proved sufficient to account for 45% of the variance in species presence in spring, 67% in summer, 72% in autumn, and 85% in winter (much higher than I expected, given the vagaries of migration, the weather and other off-site factors).

The resulting model demonstrated some interesting effects: scrape area had a highly significant effect on the number of species present, but on the same scale scrub area did not. We could therefore afford to sacrifice scrub for scrape in the interests of maximising species diversity.

It indicated that redshank presence could be enhanced more effectively by clearing surrounding scrub than by expanding the scrapes – even though redshank uses scrapes not scrub. The model suggested that on the same scrape area, we could reduce redshank days by 30% just by planting 7% of the study area with scrub. To check, after six years with no redshank despite enlarging the scrapes, we cleared some of the scrub – and back came the redshank.

Before putting in a new scrape we used the model to estimate the likely effect on daily presence of 17 key wetland bird species (Figure 10.9). Prediction and reality showed a close correlation ($r = 0.89$ in the first year, $r = 0.92$ in the second).

On-site factors could only explain 30% of the variance in wintering snipe numbers. Snipe roosted on the reserve by day, flighting out at dusk to feed on adjacent meadows. The condition of these neighbouring meadows – derelict or grazed – determined their suitability as feeding areas and together with

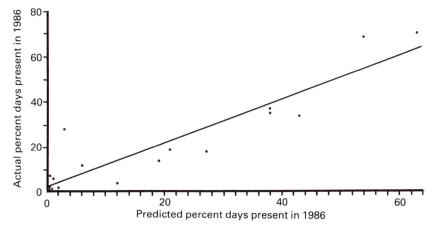

Figure 10.9 Correlation of the predicted and actual percentage of days that 17 key wetland bird species were present on the reserve in 1986, following the creation of new scrape habitats on Rye House Marsh RSPB reserve. Real data from daily bird-counts, predictions from our model based on multiple-regression analyses of on-site habitat-change and bird-species variance in the past.

on-site factors were sufficient to explain 97% of the variance in wintering populations on the reserve.

Teal numbers increased each year up to a threshold, repeating this pattern every time we extended the feeding habitat (flood-meadow). The build-up time is further evidence that tradition (new birds keep finding the site, then keep returning) is a significant but easily overlooked factor: the constancy as well as the variance of environmental factors can result in population changes.

10.16 CONCLUSION

'My overall impression is that monitoring is a lot more complex than first meets the eye and requires a lot of pre-planning and forethought' (I. Hawkins, *pers. comm.*) – that was my assistant who said that, after being coerced into reading the first draft of this chapter. But he's still monitoring, and it's still fun!

REFERENCES

Allison, J., Godwin, H. and Warren, S.H. (1952) Late glacial deposits at Nazeing in the Lee Valley. *Proceedings of the Royal Society Series B*, **236**, 169–240.

Anderson, P. and Edwards, P. (1985) *King George Reservoir, North Chingford*, Report to the Lee Valley Regional Park Authority.

Atkinson-Willes, G.L. (1963) *Wildfowl in Great Britain*, HMSO, London.

Blondel, J., Ferry, C. and Frochet, B. (1970) La méthode des indices ponctuels d'abondance (IPA) ou des relèves d'avifaune par 'stations d'ecoute', *Alauda*, **38**, 55–71.

Burnham, K.P., Anderson, D.R., Laake, J.L. (1981) Line transect estimation of bird population density using a Fourier series, in *Studies in Avian Biology No. 6* (eds C.J. Ralph and J.M. Scott), Cooper Ornithological Society, Kansas, pp. 466–82.

Groves, E.W. (1962) Vegetation of the Sanctuary, in *Bird Sanctuary* (ed. T.L. Bartlett), Selborne Society, London.

Harvey, H.J. (1982) *An ecological survey of Walthamstow Marshes*, Report to the Lee Valley Regional Park Authority.

May, R.M. (1980) Cree-Ojibwa hunting and the hare-lynx cycles, *Nature*, **286**, 108.

Mills, D.H. (1965) The distribution and food of the cormorant in Scottish inland waters, *Freshwater and salmon fisheries research*, **35**, HMSO, Edinburgh.

Muir, C. (1981) Conservation sites, *Nature*, **290**, 82–3.

Oelkel, H. (ed.) (1980) Bird census work and nature conservation. *Proc VI International conference on bird census work*. Dachverband Deutscher Avifaunisten, Lengede, Germany.

Parslow, J. (1973) *Breeding Birds of Britain and Ireland*, Poyser, Berkhamsted.

Ralph, C.J. and Scott, J.M. (eds) (1981) Estimating numbers of terrestrial birds. *Studies in Avian Biology No. 6*. Cooper Ornithological Society, Kansas.

Roberts, K.A. and Edwards, P.W. (1974) The flora of Perivale Wood Nature Reserve. *London Naturalist*, 53, 34–51.

Roberts, K.A. and White, G.J. (1985) *King George V Reservoir: effects of the proposed water sports developments on the conservation interest*, Natural Enterprises, Southampton.

Roberts, K.A. and White, G.J. (1988) The strategic duck! *B.E.S. Bulletin*, 19, 96–8.

de Schauensee, R.M. and Phelps, W.H. Jr. (1978) *A guide to the birds of Venezuela*. Princeton University Press, New Jersey.

Shenstone, J.C. (1913) The flora of the Brent Valley bird sanctuary. *Selborne Mag.* 24, 26–7, 50–1, 105–6, 146–8, 169–71, 181–5, 214–16.

Taylor, K., Fuller, R.J. and Lack, P.C. (1983) Bird Census and atlas studies. *Proceedings of the VIII International conference on bird census and atlas work.* British Trust for Ornithology, Tring.

Wurzell, B. (1980) *The Living Valley*, An open letter.

Wurzell, B. (1982) *Wild London*, 3, 3. London Wildlife Trust, London.

11

Monitoring Overseas:
Prespa National Park, Greece

BARRIE GOLDSMITH

11.1 INTRODUCTION

Ecologists and conservationists are sometimes asked to set-up monitoring schemes in countries other than their own. The response is usually positive as it is generally considered a pleasure to experience a different environment. However there can be problems especially with species identification, the assessment of the relative importance of different species and the identification of priorities for monitoring. This study deals with a national park in Greece which is known to ornithologists for its breeding population of two species of pelican and the pygmy cormorant. The status of these and other birds was of concern to several agencies including the Worldwide Fund for Nature (WWF). This study forms part of a larger one which also involves writing a management plan. It was carried out in conjunction with Greek ecologists and the Greek National Park authorities. Only the work involving monitoring is described here.

Prespa National Park is located in the north-western corner of Greece at 40°40′N and 2°32′E. It consists of parts of two lakes, a small corner of Megali Prespa and most of Mikri Prespa. Megali Prespa lies mostly in Yugoslavia and Albania but Mikri Prespa is mostly in Greece with only the southern tip in Albania (see Figure 11.1). The total area of the park is 256 km square of which 173 km square is land lying mostly to the east or the west of Mikri Prespa. Water occupies 83 km square of the park of which 48 km square is the surface of Mikri Prespa and 35 km square is Megali Prespa. The lake surface is at 853 m above sea-level and the surrounding land reaches 2156 m at the summit of Kalo Nero, a granitic area in the north-east. Much of the Park however is limestone rising to 1741 m around Triclario. Mikri Prespa is mesotrophic with a pH of around 8.1–8.5 (Mourkides *et al.* 1978). It freezes over in the winter for up to one month but reaches quite high temperatures (circa 25°C) in the summer. There are reed beds around much of the shoreline with more extensive areas along the northern isthmus which separates it from Megali Prespa and at the Albanian border. Outside the reed beds is a zone of

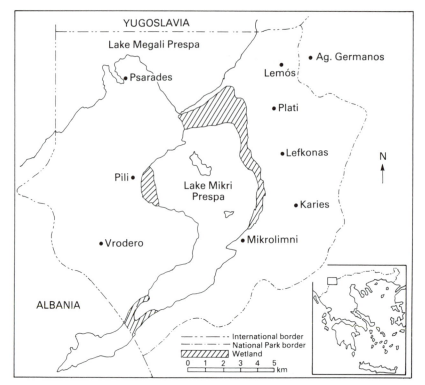

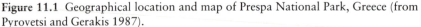

Figure 11.1 Geographical location and map of Prespa National Park, Greece (from Pyrovetsi and Gerakis 1987).

seasonally flooded grassland referred to as wet meadow which is an important feeding area for several species of birds.

The lake has many species of fish of which some are endemic (Crivelli 1984, 1989). Two species of pelican, the white and dalmation, breed in the reed beds and consume fish from the two lakes. These, and the pygmy cormorant which is quite numerous, make this a unique ornithological location. Other interesting bird species include spoonbill; grey, purple, squacco and night heron; little bittern; glossy ibis; great white and little egret; goosander (most southerly breeding location), and black-necked grebe.

The mammals of the Park include brown bear, wolf, jackal, chamois, otter and coypu. There are 1249 species of plant (Pavlides 1985) representing 111 families. These include 36 Red Data Book species and 24 endemics (see too Polunin 1987).

There are extensive oak forests on the lower slopes with *Quercus frainetto* and *Q. trojano*. These give way to beechwoods at higher altitudes, 1200–1800 m, which consist mostly of *Fagus moesiaca*. One area of beech also includes Silver fir (*Abies alba*) which is very rare in Greece. There is also an extremely important area of juniper forest, including large trees of *Juniperus excelsa* and *J. foetidissima*, bringing the total number of species of this genus to four. As well as the wet meadows around the lake there are important meadows in the forest zone and subalpine and alpine meadows where many of the endemic plants occur. A more detailed description of the Park can be obtained from Karteris and Pyrovetsi (1986), Pyrovetsi and Karteris (1986), Pyrovetsi and Crivelli (1988).

The Park has a population of about 1545 people located in ten small villages. There is some agricultural land especially where the largest stream, Agios Germanos, meets the lake where beans are an important crop. Fishing is important but the boats are mostly small and dispersed around the lake. Whilst there used to be a conflict between fishermen and fish-eating birds this is less apparent today. Some villagers keep sheep, cattle and goats but these may be important to the maintenance of alpine pastures and are only a problem on certain slopes prone to erosion. The risk of fire is also a problem and 1988 witnessed an extensive (1800 ha) and very damaging conflagration.

The WWF's interest in Mikri Prespa stems from the presence of the pygmy cormorant and the two species of pelican and their concern that excessive development could ruin the area. Recent worrying proposals have included intensive irrigation systems using water from the lake, a huge fish hatchery (built but not functioning), inappropriate tourism development, quarrying for bauxite, a hydro-electric project, more intensive farming and the use of reedbeds to filter nutrients from agricultural drainage. Fortunately the Greek National Park agency is also concerned and asked WWF to investigate five topics, one of which was to formulate a monitoring programme. In the longer term both WWF and the Greek authorities are keen to see the formulation of a comprehensive management plan for the area which will allow appropriate development whilst ensuring that the wildlife interest of the area survives. The following account sets down the key stages necessary to produce a monitoring programme for the Park and was largely funded by WWF to whom I am very grateful.

Monitoring involves repeated surveys at regular intervals of time and should only be carried out in relation to clearly defined objectives. As such it is usually aimed at recording change, or the possibility of change which may result from changes of management, exploitation, over-use or other types of environmental impact.

11.2 OBJECTIVES OF MONITORING

In Prespa National Park the objectives are to detect, record and present, in a form comprehensible to the Park authorities and the WWF, any changes, trends or impacts in the number or condition of the major biotopes or any

Table 11.1 Nature conservation evaluation of features in the Prespa National Park

		International	National	Regional
1. Biotopes	Lake	√		
	Reedswamp	√		
	Wet meadows	√		
	Fagus forest		√	
	Quercus forest		√	
	Abies forest	√		
	Juniperus	√		
	Phrygana			√
	Pseudo-steppe			√
	Alpine/Sub-alpine		√	
	Valley bottoms			√
2. Species	Dalmation pelican	√		
	White pelican	√		
	Pygmy cormorant	√		
	Great white egret	√		
	Glossy ibis		√	
	Spoonbill		√	
	Heron spp.		√	
	Goosander		√	
	Bear	√		
	Wolf	√		
	Otter		?	
	Dragonflies		?	
	Butterflies		?	

species of international, national or regional importance in the Park. These have been derived from an evaluation of wildlife features, see Table 11.1. Such monitoring should (a) serve to detect the effectiveness of any management carried out to protect or enhance any of the special features, biotopes or species of the Park; (b) be repeatable on whatever time interval is appropriate; (c) the sample points must be relocatable; (d) be sufficiently sensitive to pick up changes in a short period of time; (e) be statistically valid where appropriate; (f) be simple to carry out and not too time-consuming. Unfortunately it is difficult to satisfy all these criteria but they make a useful list of points to be considered when preparing a monitoring programme.

11.3 DECISIONS NEEDED

Having identified our objectives as primarily the monitoring of biotopes and species of international and national importance as set down in Table 11.1

and recognising that we are dealing with aquatic and terrestrial systems, as well as plants and animals which may be either sedentary or highly mobile, we must decide for each ecosystem or taxonomic group the appropriate sampling procedure. The decisions that need to be taken are to determine the optimal:

1. time interval (frequency);
2. intensity of sampling (e.g. 1%, 5%, 20%);
3. sampling unit size (e.g. number of quadrats, sites, etc.);
4. sampling pattern (grid, transect, random, stratified);
5. relocation (e.g. wooden or metal posts, coins for detection with metal detector if posts have gone);
6. time needed, time available and the best time of the year;
7. organisation of finance and administration.

Needless to say these need a considerable amount of thought and reference to Chapters 1, 2, 5 and 10 and to some standard ecological texts (e.g. Goldsmith *et al.* 1986) may be found useful.

11.4 RECOMMENDATIONS FOR MONITORING

It is clearly a complex task to draw up a programme to cover such a wide range of tasks, dealing with fish, birds, vegetation and man's exploitation of a range of biotopes from open water to sub-alpine communities. So here as discussed elsewhere in this book it is important to clarify and define our objectives. Some tasks are more important than others and there will invariably be differences of opinion between experts. Each will emphasise his or her area of specialism. However, in order to initiate the debate the various types of monitoring have been placed in three groups (see Table 11.2) ranging from 'top priority', through 'recommended' to 'optional but also desirable'.

The priorities for monitoring have been determined from the evaluation table and our knowledge of food-chains. Thus pelicans, pygmy cormorant and the great white egret are of international or national importance and feed mostly on fish. Thus monitoring should consist of direct counts of the birds, fish yields and various measures of lake water quality as a top priority. As a second priority more details of water quality should be incorporated and, if time and funds allow, lake phtyo- and zoo-plankton and dissolved oxygen should be documented.

Spoonbill, glossy ibis, greylag and waders have a more catholic diet and utilise the wet meadows more than any other single habitat. They are of national significance rather than international importance. Monitoring the extent of the wet meadows using cover-type mapping (see Chapter 3) together with data from permanent quadrats would document major changes in their habitat and direct counts would indicate changes in the numbers of

Table 11.2 Monitoring activities divided into three priority groups

	Interval
1. *Top priority*	
Fish yields	1 year
Lake transparency – secchi disk	2 weeks
Lake pH	2 weeks
Cover type mapping from air photos or satellite imagery	5 years
Breeding lake and wetland birds	1 year
Quadrats in wet meadows	2 years
Fixed point photos from viewpoints and existing towers	1 year
2. *Recommended*	
Water chemistry – phosphorus and nitrogen	1 month
Additional fixed point photos from portable towers	1 year
Forest cutting statistics	1 year
Additional water-level recording	2 weeks
3. *Optional and desirable*	
More frequent sampling of quadrats	1 year
More biotopes, more sampling points per biotope	1 year
Quadrat monitoring combined with experimentation	1 year
Lake phyto- and zoo-plankton	2 weeks
Lake dissolved oxygen	2 weeks
Experimental fish sampling	1 month
Emergence boxes for Chironomid larvae in wet meadows	May/June

these species. Quadrats need to be moved after two years to prevent the problem of autocorrelation as discussed in Chapter 2.

Bears and wolves are of international importance but are extremely elusive and occur at low densities. Direct counts would require a very large amount of time and highly skilled recorders. Therefore we do not recommend this approach although keeping a tally of known kills on roads and elsewhere would be worthwhile. We recommend documenting the extent of the key biotopes, i.e. all forest types as the top priority, together with the collation of forest cutting statistics as the second priority. The forest has national significance in terms of its silver fir (*Abies alba*) component which occurs in only one part of the National Park and is uncommon elsewhere in Greece. As a lower priority we would like to record its extent, any changes and its key components. Other major tree species in the Park are the four species of *Juniperus*, some of which (*J. foetidissima, J. excelsa*) are rare and the extensive beech (*Fagus*) forest which is of national importance.

Of the habitat types the lake is obviously vital but is not likely to change in extent, the zone of fringing reeds (*Phragmites australis*) is crucial for nesting of pelicans, spoonbill and some heron species. The area known as Vromolimni is especially important. Changes in the extent of this habitat can be recorded from fixed point photography and cover type mapping. The wet meadows are an extremely important habitat and their monitoring has already been discussed. Forest is more extensive and appears less threatened although a single large fire in 1988 has done considerable damage. If this were to become a frequent occurrence our priorities might be considerably changed. Thus monitoring is a dynamic process which must be continually reassessed. Phrygana, other scrubby vegetation and the mountain tops have

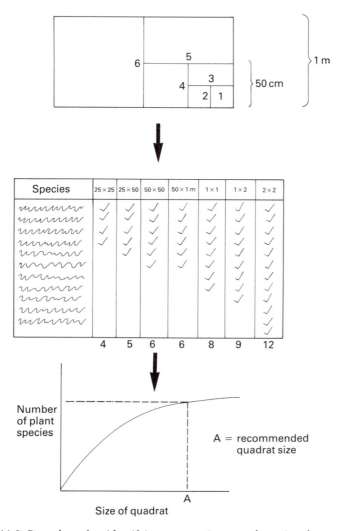

Figure 11.2 Procedure for identifying appropriate quadrat size for vegetation monitoring.

not been closely examined but contain some of the endemic plant species and presumably a characteristic suite of invertebrates. They would repay closer investigation and need monitoring based on repeated quadrat sampling.

As has been indicated the monitoring programme needs continuous review, in the light of additional information from the Park and increased knowledge of the abundance of species within and outside the Park and the extent of different habitats outside it. The evaluation of conservation importance of different features will change as will their rarity and this will continuously affect the priorities assigned to different monitoring activities.

For example the procedure for selecting an appropriate quadrat size might appear to be a mere detail but is nevertheless important because it can save time, provide more precise results and partly determines whether the results are comparable with other studies. The recommended procedure is illustrated in Figure 11.2 and was tested on the wet meadow near Mikrilimni. In practice the curve levelled off at 4×4 m but we recommend four replicates, each of 2×2 m, these being quicker and easier to record in the field. They could be used to give some estimate of variability and also permit the comparison of the results with other grassland studies where this tends to be the standard grassland quadrat (e.g. the British Nature Conservancy Council's National Vegetation Classification).

As an example of a monitoring exercise which is optional but desirable we can consider experimentation in parallel with vegetation monitoring. This was discussed with Kiriaki Kalburtji, Dimitrios Veresoglou and Sotiris Tsiouris of the Aristotelian University of Thessaloniki in the field. The discussion took place where an area of species-rich limestone grassland and scrub had the appearance of being heavily grazed (possibly from the 'nomadic' sheep from Thessaloniki). Rather than just monitor it was agreed that it would be desirable to fence one plot so that large vertebrate grazers are excluded (Figure 11.3). This could then be used to illustrate the effect of grazing which could be discussed periodically with local people. In this example monitoring could be extended to have an educational role which might have a long-term benefit to a wide range of the terrestrial resources of the Park.

Figure 11.3 also shows how two quadrat sizes may be appropriate where there are plants of different architecture. Basically a larger quadrat (20×20 m) is necessary for vegetation consisting of larger individual plants such as shrubs and a smaller one (2×2 m) for the herbaceous material consisting of grasses and herbs (see Chapter 5). In this design the large quadrats could not be easily moved but the smaller ones could to avoid the problem of autocorrelation.

Monitoring fish is notoriously difficult because most of the time they are invisible. They are highly mobile and different species show different food preferences, feeding and spawning behaviour. Experimental sampling of fish

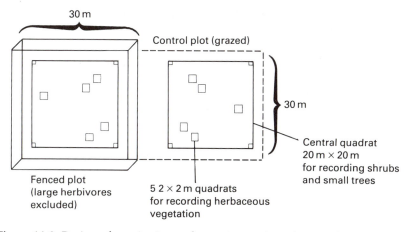

Figure 11.3 Design of monitoring and experimentation of vegetation response to change in grazing pressure.

is very time-consuming and requires large quantities of data before anything better than general trends emerge. We recommend therefore that official catch data for the main commercial species be used (see Crivelli 1984, 1989). This is likely to be an underestimate rather than an overestimate due to the 10% tax that is charged and occasionally avoided. However the 'fishing effort' is likely to be somewhat variable and it would be desirable to also record 'catch per unit effort'. If fishermen knew that such records were being kept it is possible that their fishing returns would be made more conscientiously and so the tax revenue would increase and the fish data would be more reliable.

Monitoring birds is less difficult than with fish but still poses problems. We need to focus on the internationally important species such as the two species of pelican and pygmy cormorant. The breeding season is the best time and it would be desirable if records could include information about behaviour and breeding. The two main areas for study should be Vromolimni and the surrounding reed swamp and the area at the southern end of the lake near the Albanian border. Both species of pelican and the pygmy cormorant are the top priority followed by spoonbill, great white egret and heron species and then information about the areas visited and duration of stay of glossy ibis. If the person doing the work is an ornithologist there would be real merit in point counts at other stations, including the observation tower. Where possible information about the areas used for feeding, 'loafing', and other activities should also be recorded. At a lower level of priority additional counts could be taken from the portable towers at times when they are in use.

To record the 'general condition' of agricultural land and any deterioration that might have occurred due to the excessive use of inorganic fertiliser, pesticides or excessive leaching from irrigation is a daunting task. One procedure which is simple, relatively easy and can be enjoyable is to follow the procedure of Pollard (1982 and Chapter 6 in this volume) which involves a standard walk, divided into sections involving several different habitats, recording the numbers of each butterfly species on each section. The sections can be of variable length but should be a standard width and height such as

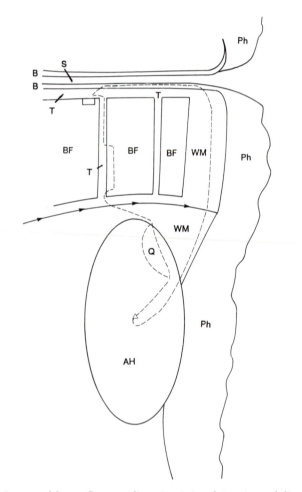

Figure 11.4 Suggested butterfly recording circuit involving Amygdalies Hill (AH). BF = bean field, B = bank, S = seasonal stream, Ph = *Phragmites*, WM = wet meadow, Q = quarry.

5 m. This needs to be carried out at a standard time (say between noon and 3 pm) on sunny, warm days. It has proved particularly effective in the UK and, as 27 species were recorded in casual observation in the Prespa National Park in two weeks at the end of October and in May, should have considerable potential. The results would indicate the differences between habitats, time of year and annual trends of 40 or more species which are readily identifiable by a good all-round field biologist. A suitable area would be Amygdalies Hill, 903 m, which is limestone, set in very intensive arable land and closely adjacent to wet meadow (see Figure 11.4). The suggested walk would cover five different habitat types in the duration of a half-hour walk. If similar data were collected in other places in Greece useful comparisons could also be made to indicate effects of altitude, latitude, surrounding land-use, etc.

11.5 STAFFING, COSTS AND ADMINISTRATION

Monitoring is such an important activity that it requires a full-time worker throughout the year. He or she could be called the monitoring warden and would need a scientific training to degree and preferably post-graduate level. The activities would be distributed though the year as shown in Table 11.3.

The cost of employing this person would be a minimum of £8000 plus 35% overheads plus a contribution to the cost of running a car (at least £1000 per annum) plus a contingency sum to cover the cost of laboratory analysis of lake samples if required (£1000 per annum) plus the cost of visiting experts for advice if required (say £1000 per annum). A grand total of at least £13 000 per annum. This person could also be expected to help with wardening the Park, reporting infringements of the law to the 'agricultural policeman', etc. and help with the provision of interpretive material for visitor centres and schools. Thus monitoring is not cheap and it might be that an international agency would need to help to ensure that the programme became established.

Table 11.3 Times of year for various monitoring activities

Monitoring activity	Month											
	J	F	M	A	M	J	J	A	S	O	N	D
Lake chemistry and fish	√	√	√	√	√	√	√	√	√	√	√	√
Vegetation quadrats						√'	√	√				
Cover type mapping	√	√								√	√	√
Birds				√'	√	√			√'	√'		
Forest cutting statistics	√	√								√	√	√

√ = top priority.
√' = other recommended times depending upon workload and priorities.

11.6 CONCLUSION

If the intensity of monitoring recommended was carried out, i.e. categories 1 and 2 at least, it would ensure that changes in any major biotope or important species were detected early and remedial action could be implemented. In addition important information about the biological resources of the Park would be documented, publications could be prepared and the 'Biological Warden' could help with the preparation of interpretive material. A feedback mechanism should operate whereby results of monitoring influence the 'prescription' section of the management plan which itself should evolve as more information about ecological trends in the Park are documented. Monitoring will also help clarify the 'specified limits' of key taxa and biotopes in the area.

REFERENCES

Crivelli, A.J. (1984) 'Lakes-Fisheries', in *Integrated Environmental Study of Prespa National Park, Greece,* (eds Pyrovetsi *et al.*). Final Report, Commission of the European Communities DG XI, 49–86 and 144–55.

Crivelli, A.J. (1989) Fisheries decline in the freshwater lakes of northern Greece. A case study: Lake Mikri Prespa. *Proceedings of the FAO Symposium on Management schemes for inland fisheries.* Goteborg, May 1988.

Goldsmith, F.B., Harrison, C.M. and Morton, A.J. (1986) Description and Analysis of Vegetation, in *Methods in Plant Ecology* (ed. P.D. Moore), Blackwell, Oxford.

Karteris, M. and Pyrovesti, M. (1986) Land cover use analysis of Prespa National Park, Greece, *Environmental Conservation,* 13, 319–30.

Mourkides, G.A., Tsikritsis, G.E., Tsouris, S.E. and Menkisoglou, U. (1978) The Lakes of Northern Greece I. Trophic status 1977, *Scientific Annals of Section of Agriculture*, The University of Thessaloniki, Greece, 21, (5).

Pollard, E. (1982) Monitoring butterfly abundance in relation to the management of a nature reserve. *Biological Conservation,* 24, 317–28.

Pavlides, G. (1985) *Geobotanical study of the National Park of Lakes Prespa (N.W. Greece) Part A: Ecology, Flora, Phytogeography, Vegetation.* Aristotelian University of Thessaloniki, Greece (in Greek).

Polunin, O. (1987) *Flowers of Greece and the Balkans: A Field Guide.* Oxford University Press, Oxford.

Pyrovetsi, M.D. and Crivelli, A.J. (1988) Habitat Use by Water-birds in Prespa National Park, Greece, *Biological Conservation,* 45, 135–53.

Pyrovetsi, M.D. and Gerakis, P.A. (1987) Environmental problems from practising agriculture in Prespa National Park, Greece, *Environmentalist,* 7(1), 35–42.

Pyrovetsi, M.D. and Karteris, M.A. (1986) Forty year Land Cover/Use Changes in Prespa National Park, Greece, *Journal of Environmental Management,* 23, 173–83.

- 12

The calculation of index numbers from wildlife monitoring data

TERENCE J. CRAWFORD

12.1 INTRODUCTION

In the UK, on the Tuesday nearest to the middle of each month, a major shopping expedition is undertaken – but it is an unusual one in that nothing is purchased. On these occasions Administrative Grade Civil Servants from 180 of the nation's Unemployment Benefit Offices each visit a large number (up to 100) of retail outlets and record the current prices of a wide range of commodities, from bags of flour to boxes of matches and from petrol to pairs of Wellington boots. These data, together with information on other items of household expenditure, for example car insurance and telephone rentals, are used to compile the monthly Retail Prices Index (RPI). The RPI aims to provide an indication of relative price movements across a typical entire household expenditure; it is not an absolute measure of the cost of living, though from 1914 to 1947 its predecessors were known as the Cost of Living Index. Readers who would like to know more about the details of how the RPI is computed are referred to Department of Employment and Productivity (1967).

The RPI has become in recent years a familiar statistic for the population at large, not least because of its relevance to wage-bargaining or to index-related pensions. Huge sums of government expediture are linked to the RPI and it is, therefore, very important that this series of figures should provide an accurate indication of price movements. The index has a history of refinement and there remain problems in its construction that are yet to be resolved (Fry and Pashardes 1986).

In a manner somewhat analogous to our Civil Servant 'shoppers', lepidopterists make weekly counts of butterflies from April to September along transects at over 80 sites throughout the UK; this is the Butterfly Monitoring Scheme described in Chapter 6. Large numbers of amateur ornithologists participate in the monitoring schemes co-ordinated by the British Trust for Ornithology and the Wildfowl and Wetlands Trust (see Chapter 7). These schemes generate annual indices for butterflies and for many species of birds with a view to highlighting changes in their relative abundances over time. There are other monitoring programmes in the UK that do not currently lead to indices but do generate suitable data, for example the moth-trap data of the Rothamsted Insect Survey.

Just as the RPI takes account of the prices for many commodities in different towns, so also do the wildlife monitoring schemes collect abundance data for different species at many sites. However, whereas the RPI combines the information from all commodities into a single index, this is not normally so for wildlife indices which are usually compiled separately for each species. In constructing the RPI each commodity is weighted relative to its consumption. There is no analogous quantitative criterion on which data for different species can be combined; any multi-species indices must contain an element of subjectivity.

The purpose of this chapter is to describe the construction of indices from wildlife monitoring data and to consider some of the problems peculiar to biological data of these types that are not shared by economic data.

12.2 INDEX NUMBERS AND THEIR PROPERTIES

Index numbers, sometimes simply called indices, are used to measure changes between different circumstances in the values of some quantity or quantities. The circumstances may be temporal, for example a series of annual index numbers, and as this is often the case with wildlife monitoring index numbers we shall concentrate on a temporal context. There is no reason, however, why index numbers should not be used in other contexts; spatial index numbers, for example, can be used to compare different regions. Whatever the context, it is essential that comparable techniques are used on each occasion that observations are made.

Strictly speaking, for a series of values to be called index numbers they must be scaled relative to the value obtained at one particular time. This is known as the *reference base* of the series and conventionally it is assigned a value of 100. To illustrate this, let us consider the annual numbers of a certain species observed from 1980 to 1986 (Table 2.1).

In the first series the year 1980 is taken as the reference base. Each year's index number is obtained as the ratio between the number of individuals observed in the year to that in 1980, multiplied by 100. Thus, for 1984 we obtain $(9920/5381) \times 100 = 184.4$. This is equivalent to saying that abundance in 1984 was 84.4% greater than that of the reference base year, in this case 1980. The second series takes abundance in 1984 as the reference base; this time index numbers are obtained from ratios relative to the 9920 individuals observed in 1984. Note that the second series can be simply obtained by multiplying the first series by the ratio of abundances in 1980 to 1984 ($5381/9920 = 0.542$). Indeed, the second series can be obtained from

Table 12.1 A series of annual species abundances and corresponding index numbers using two different reference bases

Year	1980	1981	1982	1983	1984	1985	1986
Nos obs.	5381	4820	6421	7715	9920	8147	7352
1980 = 100	100	89.6	119.3	143.4	184.4	151.4	136.6
1984 = 100	54.2	48.6	64.7	77.8	100	82.1	74.1

the first without reference to the actual abundances by multiplying by the ratio of the two index numbers (100/184.4 = 0.542). The abundance in 1980 is 54.2% that of 1984.

Do not be perturbed by the fact that when 1984 is the reference base the 1980 index number is 45.8 points less than that of 1984, whereas when 1980 is the reference base the 1984 index number is 84.4 points greater than that of 1980. The two series really are the same. An increase of 84.4% in one direction is the same as a decrease of 45.8% in the other direction; this is the way that proportions, percentages and index numbers work. It is like saying that if x is twice y then, expressed the other way around, y is one-half of x; one is the reciprocal of the other. In the same way, our index numbers are reciprocals of one another, i.e. $1.844 \times 0.542 = 1.0$. The real point is that index numbers are relative, rather than absolute, values. This has important consequences that are best illustrated graphically. In Figure 12.1(a) the two series are plotted on a linear scale. As expected, the two lines are always vertically separated. The line based on 1980 gives an apparently optimistic series of index numbers, only one being less than 100. By contrast, the index numbers based on 1984, a more prolific year, seem in general pessimistic, the highest being the value of 100 for 1984 itself. Technically, these differences are immaterial when we remember that the index numbers depend only on abundances relative to those of the reference base year. Year-to-year comparisons of index numbers reflect changes in abundance; they convey no information at all about the actual abundance in any particular year. In practice, however, because not everybody who comes across index numbers appreciates this point, there is something to be said for picking a reasonably typical year as the reference base.

A more important difference between the lines in Figure 12.1(a) is that their detailed shapes are different. If the reference base year has an atypically low value, differences between index numbers in the series become exaggerated, whereas if it has a high value, variation in the series is damped. The variance of the series changes inversely with the square of the ratio of abundances, or index numbers, in the two reference base years. In the present example, the ratio for 1980 to 1984 is 0.542 and the variances of the series based on 1980 and 1984 are respectively 1042.1 and 306.6; $1042.1/306.6 = 3.4 = 1/0.542^2$. Thus, the fluctuation in a series of index numbers will depend upon the

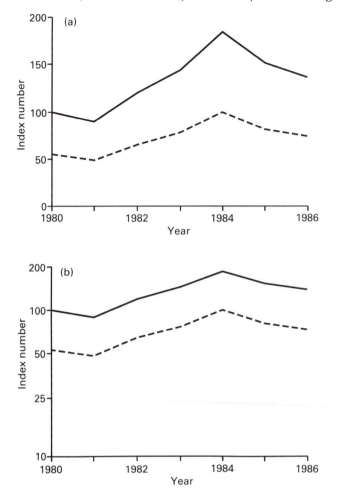

Figure 12.1 Series of index numbers calculated from the same run of data plotted on (a) a linear scale and (b) a logarithmic scale: —— 1980 and ――― 1984 taken as reference base.

reference base chosen and this can lead to problems of interpretation, particularly when series from different sets of data, say different species, are to be compared. The solution is simple; the series should be plotted on a logarithmic scale, as in Figure 12.1(b). The shape of the series is then the same whatever reference base is adopted and is simply shifted vertically, upwards for a low reference base, and vice-versa.

Because index numbers are ratios they have no dimensions. This is a very useful property because it means that relative changes in a variety of variables, each with quite different dimensions, can be combined with one another into a composite index number. It was in this context that index numbers were originally developed to measure changes in aspects of the economy which could not be measured directly. For example, it is impossible to observe and measure directly changes in the value of money in the sense that we can with the rate of a biochemical reaction or the number of mallards on a given pond on different dates. The RPI is an attempt to deal with the value of money problem by combining information on price relatives for a representative 'shopping basket' of a variety of commodities. To a purist, therefore, the index numbers we have calculated in our species abundance example are not, strictly, index numbers; the abundances are directly measurable and all we have done is to re-scale them. To call them index numbers is not, however, at variance with popular practice in the context of wildlife monitoring data. Indeed, the term is sometimes applied to abundances that have not been scaled relative to a reference base with a value of 100 – see the Butterfly Monitoring Scheme (Chapter 6).

Let us now consider a slightly more realistic example. Suppose we have species abundance data for five sites measured in each of ten years. The data are given in Table 12.2. I have made them simple on purpose: no missing observations and no zero observations. In the calculations that follow, it may sometimes seem that an unnecessary degree of accuracy has been employed. This is to aid technical comparisons and because it is always sensible in intermediate calculations of this type to carry a high number of significant figures and to round at the end.

The essence of the problem is that a series of index numbers must be found that can reflect changes in abundance at all five sites in a fair way; no one site should influence the series more than the other sites. The obvious initial approach is simply to sum abundances over sites for each year and to use the summed abundances to compute a series of index numbers relative to a chosen reference base year. Table 12.2 shows series of index numbers using both years 1 and 5 as reference bases. Note that the two series are the same; the second can be obtained from the first simply by multiplying by $100/253.33 = 0.3947$. In all that follows this important relationship between series based on different reference base years will be assumed to hold. Unless stated otherwise, from now on index numbers will be based on year 1, and it may be taken that series based on other years can be obtained by multiplying by the ratio of the index number in the old reference base year to that in the new.

Although straightforward, and superficially sensible, this method has its limitations. The changes in abundance within any individual site lose their integrity; an index number reflects changes summed over all sites irrespective of the local significance of those individual changes. In Table 12.2 the

Table 12.2 Fictional data for species abundance at five sites in each of ten years

Year	Site					Sum	I_1	I_5
	1	2	3	4	5			
1	3	10	7	8	2	30	100.00	39.47
2	5	15	4	3	1	28	93.33	36.84
3	3	16	6	10	1	36	120.00	47.37
4	12	41	10	22	6	91	303.33	119.74
5	6	27	11	28	4	76	253.33	100.00
6	2	18	4	12	2	38	126.67	50.00
7	1	9	7	9	1	27	90.00	35.53
8	1	5	4	2	1	13	43.33	17.11
9	2	7	7	9	3	28	93.33	36.84
10	1	7	9	10	3	30	100.00	39.47

I_1 and I_5 are index numbers with years 1 and 5 as reference base.

summed abundances are the same in year 10 as in year 1 and, therefore, the index numbers in year 10 are the same as those in year 1. But there is a variety of ways by which a total abundance of 30 could have been achieved in year 10. What if four sites had lower abundances in year 10 compared with year 1, but the fifth site had an increased abundance that was sufficiently large to compensate for the decreases in the other four? The index numbers would be the same, but is it fair that the changes at one site should be allowed to mask changes in the opposite direction at *four* other sites?

A possible way around this problem would be to compute index numbers separately for each site, relative to a common reference base year, and then to use the sum of the individual site index numbers as the basis of a composite index number for all five sites. The calculations are shown in Table 12.2 where year 1 is used as the reference base for each site and also for the composite index numbers. For example, the index number for site 2 in year 8 is (from Table 12.2) (5/10) × 100 = 50. Comparison between Tables 12.3 and 12.4 shows that the series of composite annual index numbers does differ from the index numbers obtained by simple summation of abundances across sites, and no longer is the final index number in the series equal to the first. The differences arise because, with the first method, changes in unproductive sites contribute less to the index numbers than do similar relative changes in the more productive sites. In this respect the second method would seem to be an improvement. In year 5, for example, both sites 1 and 5, where abundances doubled compared with year 1, now make an equal contribution to the composite index numbers.

There is, however, a problem with this method that is not readily apparent.

Table 12.3 Site and composite index numbers calculated from the data in Table 12.2

Year	Site index numbers					Sum	Composite index nos
	1	2	3	4	5		
1	100.00	100.00	100.00	100.00	100.00	500.00	100.00
2	166.67	150.00	57.14	37.50	50.00	461.13	92.26
3	100.00	160.00	85.71	125.00	50.00	520.71	104.14
4	400.00	410.00	142.86	275.00	300.00	1527.86	305.57
5	200.00	270.00	157.14	350.00	200.00	1177.14	235.43
6	66.67	180.00	57.14	150.00	100.00	553.81	110.76
7	33.33	90.00	100.00	112.50	50.00	385.83	77.17
8	33.33	50.00	57.14	25.00	50.00	215.47	43.09
9	66.67	70.00	100.00	112.50	150.00	499.17	99.83
10	33.33	70.00	128.57	125.00	150.00	506.90	101.38

Year 1 is reference base for site and composite index numbers.

The calculations of Table 12.3 are repeated in Table 12.4, but this time year 5 is taken as the reference base for the site index numbers and year 1, as before, for the composite index numbers. Now a series is generated that again differs from that of Table 12.2, but it also differs from that of Table 12.3! Adopting year 5 as the reference base for the composite index numbers does not help; in relative terms the composite series remains the same. The problem is that the series of composite index numbers will differ according to which reference

Table 12.4 Site and composite index numbers calculated from the data in Table 12.2

Year	Site index numbers					Sum	Composite index nos
	1	2	3	4	5		
1	50.00	37.04	63.64	28.57	50.00	229.25	100.00
2	83.33	55.55	36.36	107.10	25.00	210.95	92.02
3	50.00	59.26	54.55	35.71	25.00	224.52	97.94
4	200.00	151.85	90.91	78.57	150.00	671.33	292.84
5	100.00	100.00	100.00	100.00	100.00	500.00	218.10
6	33.33	66.67	36.36	42.86	50.00	229.22	99.99
7	16.67	33.33	63.64	32.14	25.00	170.78	74.50
8	16.67	18.52	36.36	7.14	25.00	103.69	45.23
9	33.33	25.93	63.64	32.14	75.00	230.04	100.34
10	16.67	25.93	81.82	35.71	75.00	235.13	102.56

Year 5 is reference base for site index numbers and year 1 for composite index numbers.

base year is adopted for the site index numbers. The individual site series are comparable, but as they are ratios their sums are scale-dependent. Consider just two sites and two years, where numbers, whatever they are, double in one site and quadruple in the other. Taking the first year as reference base for each site, the sums of the site index numbers are 200 (100 + 100) for the first year and 600 (200 + 400) for the second year producing composite index numbers of 100 and 300. If, instead, the second year is taken as the reference base for site index numbers, their sums become 75 (50 + 25) and 200 (100 + 100) leading instead to composite index numbers of 100 and 267.

Given that the composite series depends upon the choice of reference base year for the individual sites, it is reasonable to enquire whether it is possible to find an optimum reference base. Because, for any single site, a series of index numbers is being used, rather than raw abundance data, it follows that the absolute values, but not the relative values, in the series are dependent on the reference base (see Tables 12.3 and 12.4). Rather than using a common reference base year for all sites it may be preferable to adopt different ones, picking for each site a representative year. The figure taken as the reference base does not need to be an abundance scored in a particular year – any figure at all will generate the same *relative* movements in the series for an individual site. A potentially attractive figure, and certainly a representative one, is the mean abundance for each site, i.e. use 3.6 as reference base for site 1, 15.5 for site 2, etc. For comparison with the results of earlier calculations, the composite index numbers then become:

Year:	1	2	3	4	5	6	7
Index No:	100	89.69	100.06	294.34	226.74	103.90	76.49

Year:	8	9	10
Index No:	43.97	100.84	103.33

As more annual data are collected the site means will change. But these changes will usually become smaller as time passes. For example, Taylor *et al.* (1981) considered the annual catches between 1969 and 1978 of 20 species of moths at 15 Rothamsted light-traps, five in each of three regions (southern England, Wales with central and northern England, and Scotland). Within each region, and for all 15 traps, they calculated progressive annual arithmetic means for each species by adding in successive year's data, i.e. 1969 alone, mean of 1969 and 1970, mean of 1969, 1970 and 1971, etc. For all species except two, these means quickly settled down to a reasonably constant value, often after only three years.

The imaginary data in Table 12.2 intentionally have a feature that is common to many natural populations. Larger populations often show more variation in population size, and more common species show more variability in both spatial and temporal abundances. In these cases variation in population size is better described on a logarithmic scale than on a linear one.

Table 12.5 Index numbers calculated from the data in Table 12.2 after logarithmic transformation

Year	Site 1	2	3	4	5	Sum	Index nos	G.M.	Index nos
1	0.477	1.000	0.845	0.903	0.301	3.526	100.00	5.073	100.00
2	0.699	1.176	0.602	0.477	0.000	2.954	83.78	3.898	76.84
3	0.477	1.204	0.778	1.000	0.000	3.459	98.10	4.919	96.96
4	1.079	1.613	1.000	1.342	0.778	5.813	164.86	14.539	286.60
5	0.778	1.431	1.041	1.447	0.602	5.300	150.31	11.483	226.36
6	0.301	1.255	0.602	1.079	0.301	3.539	100.37	5.102	100.57
7	0.000	0.954	0.845	0.954	0.000	2.754	78.11	3.554	70.06
8	0.000	0.699	0.602	0.301	0.000	1.602	45.43	2.091	41.22
9	0.301	0.845	0.845	0.954	0.477	3.423	97.08	4.836	95.33
10	0.000	0.845	0.954	1.000	0.477	3.276	92.91	4.521	89.12

Year 1 is the reference base. G.M.: geometric mean.

Over a wide range of species the standard deviation of the logarithm of population size is surprisingly constant; Williamson (1972, Chapter 1) cites a number of examples. It is of interest, therefore, to investigate index numbers calculated from the site abundances after logarithmic transformation. It is common practice to add a constant, usually 1, to the abundances before transformation, otherwise zero observations cannot be transformed. As there are no zero observations in the data in Table 12.2, no constant has been added so that a more direct comparison can be made with previous results.

Table 12.5 contains the transformed data and, in the first column of index numbers, the series obtained from annual summation across sites, i.e. the transformed version of Table 12.2. The main consequence of the transformation, as expected, is to reduce the variation in the index numbers around the value of 100; in particular, very high index numbers are considerably reduced. Whether or not this is an advantage is debatable. High index numbers may often reflect temporary upward fluctuations on the underlying logarithmic scale of variability in species abundance and these fluctuations are over-emphasised on a linear scale. They might then be more properly dealt with by the logarithmic transformation, in much the same way that biometrical geneticists regularly transform continuously varying traits to a scale in accord with the underlying gene action, and on which predictions can be made (Mather and Jinks 1982). On the other hand, the potential damping effect on very low index numbers might be undesirable. Given that a primary aim of wildlife monitoring is to detect declines in species adbundance, the logarithmic transformation as employed here is quite unsuitable; once abundances were significantly below those in the reference base year, further chronic declines would become somewhat less noticeable in index numbers based on log.-transformed data. Composite index numbers, following the methods of Tables 12.3 and 12.4, but using transformed data, show no advantages compared with those based on untransformed data; the dependence on choice of reference base year is still a problem.

The analysis in Table 12.5 is taken a little further. The sums of the log. abundances in each year are divided by 5 to give a mean which is then antilogged. This produces a statistic called the *geometric mean* of the site abundances in a year. Whereas an arithmetic mean is calculated as $\Sigma\ x/n$, a geometric mean is calculated as $(\Pi\ x)^{1/n}$, i.e. the nth root of the *product* of the sample of n observations. $(x_1.x_2.\ ...\ .x_n)^{1/n}$ can be equally calculated as antilog$[(\log x_1 + \log x_2 + ... + \log x_n)/n]$, as in Table 12.5. A series of index numbers is generated that clearly belongs to the same family as those in Tables 12.2, 12.3 and 12.4. The higher values sit very comfortably amongst those calculated from untransformed data; small index numbers are marginally reduced which, because they will better indicate chronic declines, is probably not a serious disadvantage. Index numbers based on geometric means have, however, a particularly satisying property. If series of index numbers are calculated for individual sites as in Tables 12.3 and 12.4, composite index numbers compiled, not from site sums, but from the geometric means for each year, the resulting series are *exactly* the same as that in Table 12.5. This is true whichever year is chosen as the reference base for the site series, if different reference base years are chosen for individual sites, or even if quite arbitrary and different numbers are chosen as reference bases for each site. If you don't believe this, try it for yourself! It would seem that index numbers based on geometric means, i.e. arithmetic averages of logarithmically transformed abundances, are very well-behaved. Note that the base of the logarithms is immaterial; common and natural logarithms give exactly the same results. It should be added that geometric means hold a very respectable position in the history of the construction of economic index numbers.

12.3 WILDLIFE INDEX NUMBERS IN PRACTICE

The data that we have so far considered are ideal in the sense that every site is represented in every year. It may be possible to achieve this for small-scale monitoring programmes with sites carefully chosen for their potential longevity, or for monitoring rare species. It can be argued, however, that our general wildlife resource deserves monitoring attention; attrition in more common and widespread species should be a matter of great concern, not least because they are unlikely to enjoy the benefits of positive conservation measures. They might, therefore, be better indicator species for the general health of our wildlife. Ideally, such species should be monitored on a national scale at a reasonably large number of sites. This has been achieved by the bird-monitoring schemes of the British Trust for Ornithology (see Chapter 7) and the Wildfowl and Wetlands Trust, the Butterfly Monitoring Scheme of the Institute of Terrestrial Ecology (Chapter 6) and for moths by the Rothamsted Insect Survey.

It goes without saying that these large-scale national schemes depend to a considerable degree upon the efforts of volunteers. This immediately introduces the problem that volunteers cannot be relied upon to produce a continuity of data over a long period of time. Individual enthusiasms wax and wane, or people move; at worst, they may die. These days we cannot even count on the continued existence of well-established research institutes which might, for example, operate a moth trap on their site. The result is that discontinuities and holes appear in the sites times years data matrix. This has been the experience of all the schemes mentioned above.

The British Trust for Ornithology approached the problem in the following way. In any one year, only those sites are included that also contributed data in the previous year. The species abundances are summed for the paired sites in the two years. If the reference base year is the first year in the series, the calculations are very simple. The first year is assigned an index number of 100. The second year's index number is obtained by multiplying that of the first year (100) by the ratio of the total abundances, summed over the paired sites, of the second year to those of the first. Similarly, to calculate the index number for the third year, a ratio is obtained from the summed abundances of sites paired between the third and second years; the new index number is obtained by multiplying that of the second year by this ratio. Table 12.6

Table 12.6 Data of Table 12.2 with some missing observations: used to calculate index numbers in Table 12.7

| Year | Site | | | | | Sum |
	1	2	3	4	5	
1	3	10	7	8	–	28
2	5	15	4	3	–	27
3	–	16	6	10	1	33
4	12	41	–	22	6	81
5	6	27	–	28	4	65
6	2	18	4	12	2	38
7	1	9	7	9	1	27
8	1	5	4	2	–	12
9	–	7	7	9	3	26
10	1	7	9	10	3	30

Table 12.7 The calculation of index numbers by the paired sites method when the sites times years matrix is incomplete

Year	Paired-site abundances Previous year	Current year	Ratio	I_1	Ratio	I_5
1			—	100.0	1.037	34.8
2	28	27	0.964	96.4	0.688	33.5
3	22	32	1.455	140.3	0.391	48.8
4	27	69	2.556	358.4	1.246	124.6
5	81	65	0.803	287.6	—	100.0
6	65	34	0.523	150.5	0.523	52.3
7	38	27	0.711	106.9	0.711	37.2
8	26	12	0.462	49.3	0.462	17.2
9	11	23	2.091	103.2	2.091	35.9
10	26	29	1.115	115.1	1.115	40.0

I_1 and I_5 are index numbers with years 1 and 5 as reference base.

repeats the data of Table 12.2, but with some missing observations, and the method is illustrated in Table 12.7, firstly with year 1 as the reference base. Note that the calculations were performed to a degree of accuracy greater than that shown in Table 12.7; a slight drift in the series will occur if the ratios given in the table are used.

A little more care is required if the reference base is not the start of the series. In Table 12.7, year 5 is taken as an alternative reference base, as in Table 12.2. An index number of 100 is assigned to year 5 and subsequent years' index numbers are calculated as above. To work backwards from the reference base year, the previous year's index number is obtained by multiplying by the ratio of summed paired-site abundances in that year to those in the reference base year, and so on. Note that the ratio is the other way around before the reference base year. Thus, the index number for year 4 is $(81/65) \times 100 = 124.62$ and that for year 3 is $(27/69) \times 124.62 = 48.76$.

Comparison of the series of index numbers in Table 12.7 with those in Table 12.2 does show differences. As will be demonstrated later, perturbations persist in the series when index numbers are computed by this ratio method with missing observations. Over time, perturbations in one direction may compensate for those in the other direction. In any event, the problem is bound to be particularly acute in this example because of the very small number of sites. It should be noted that the ratio method applied to a complete dataset will yield exactly the same series as the method of Table 12.2; the problem lies in the missing observations rather than in the method itself.

It is useful to employ a genuine set of data to investigate further the behaviour of index numbers computed by the ratio method. The moth-trap records of the Rothamsted Insect Survey provide an ideal source of data. Taylor (1986) gives an overview of the history, work and results of the Survey, which involves aphids as well as moths. For moths, a large network of light-traps has been operated for over 20 years. The annual abundances for 1965 to 1986 of 31 moth species of economic or migratory interest have been summarised in the annual *Rothamsted Report* for the years 1968 to 1987. Tables provide the total counts for each species for each trap that completed 364 night samples (the nights beginning 31 December and 29 February are excluded). These tables are ideal for our present purposes because, not only are abundances given for the current year, but also for the previous year indicating those traps which were not operational or which failed to complete the 364 night samples. The calculation of paired-site abundances between successive years is, therefore, greatly facilitated. A little care is sometimes required, however, when interpreting these tables; trap-sites sometimes change reference numbers and I have used my own judgement as to whether or not they should be considered as the same site. There are also a few instances where data for a site are not included for comparison in the next year's table. Others working with these data may not, therefore, reproduce exactly my summarisation, but differences will be only marginal. Some sites are regularly present, others appear for shorter parts of the series and still others are sporadic; there are plenty of 'holes' in the sites times years dataset. Attention will be concentrated on the heart and dart moth, *Agrotis exclamationis*, although the magpie moth, *Abraxas grossulariata*, will also be used for comparison in one instance.

The paired-site sums for the heart and dart moth are given in Table 12.8, together with index numbers taking 1965 as the reference base year. Although it doesn't really matter for our present purposes, 1965 would not be an ideal year to choose as reference base because there are only data for seven sites; all seven were represented in 1966, but they are sites in southern England where abundances are generally higher than for sites further north.

Figure 12.2(a) shows the series of index numbers for both the heart and dart moth and also the magpie moth, taking for both species 1965 as the reference base year. The immediate impression is that, apart from 1976 and 1977 when the heart and dart moth was exceptionally abundant, both species show similar levels of abundance and degrees of fluctuation. Figure 12.2(b) shows the series recalculated with 1976 as the reference base year. The impression now is entirely different. The magpie moth would seem very much more abundant and subject to more wild fluctuations. The impressions gained with respect to differences in abundance of the two species are entirely spurious; as stressed before, index numbers relate purely to abundance relative to the reference base year for a given species. Either presentation in

Table 12.8 Index numbers for the heart and dart moth calculated by the paired-sites method

Year	Paired-site abundances Previous year	Current year	No. of sites	Ratio	I_{65}	Ratio	I_{76}
1965				—	100	1.58	24
1966	202	128	7	0.63	63	0.88	15
1967	491	561	28	1.14	72	0.44	17
1968	535	1225	28	2.29	164	3.06	38
1969	2378	778	38	0.33	54	0.29	12
1970	949	3297	45	3.47	188	1.01	43
1971	5459	5381	57	0.99	186	1.74	43
1972	7784	4480	72	0.58	108	1.50	25
1973	4014	2682	64	0.67	72	0.85	16
1974	3036	3558	65	1.17	85	1.15	19
1975	4407	3848	80	0.87	74	0.17	17
1976	3704	22098	87	5.97	440	—	100
1977	31456	25659	93	0.82	361	0.82	82
1978	21586	6093	73	0.28	101	0.28	23
1979	5459	1523	66	0.28	28	0.28	6
1980	1336	2793	63	2.09	59	2.09	13
1981	3582	1046	79	0.29	17	0.29	4
1982	991	3593	74	3.63	62	3.63	14
1983	3504	1739	69	0.50	31	0.50	7
1984	1480	7226	56	4.88	152	4.88	35
1985	6978	7700	53	1.10	167	1.10	38
1986	8453	1357	63	0.16	27	0.16	6

I_{65} and I_{76} are index numbers with years 1965 and 1976 as reference base.

Figure 12.2 simply shows that the heart and dart moth was about 4.5 times more abundant in 1976 than it was in 1965, whereas the magpie moth was about 2.5 times more abundant in 1965 compared with 1976. Of the two species, the heart and dart moth is generally very much more numerous in absolute terms. More troublesome are the differences between the species in their year-to-year fluctuations which are dependent upon choice of reference base year. It has already been suggested that plotting the series on a logarithmic scale will solve the problem, and this example provides a clear illustration. The series calculated from the same two reference base years are shown in Figures 12.3(a) and 12.3(b) plotted on a logarithmic scale. The vertical displacements of the series simply reflect relative abundances in the year chosen as reference base, a good year producing a downwards

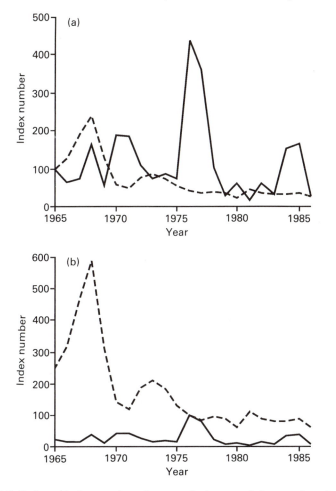

Figure 12.2 Series of index numbers for —— the heart and dart moth and − − − the magpie moth plotted on a linear scale; reference base year is (a) 1965 and (b) 1976.

displacement and vice versa. The important thing to note is that the shapes of the series are now independent of reference base year. It is clear that the heart and dart moth is subject to more violent fluctuations between years than is the magpie moth. Although the two series suggest that both species have declined in abundance to an almost equal extent during the 21-year period, it would seem that in the case of the magpie moth this reflects a continued chronic decline, whereas for the heart and dart moth it could easily be part of its naturally large fluctuations in abundance. Incidentally, it is encouraging to note that the heart and dart and magpie moth series in Figures 12.3(a) and

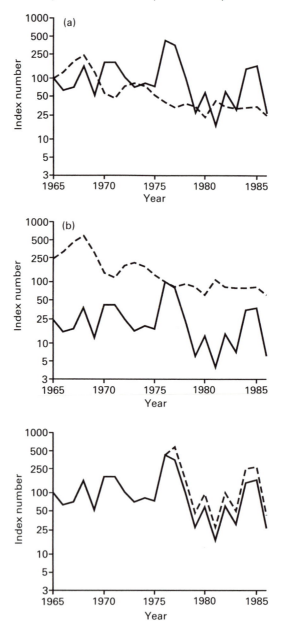

Figure 12.3 (a) and (b) Series of index numbers for —— the heart and dart moth and
——— the magpie moth plotted on a logarithmic scale; reference base year is (a) 1965
and (b) 1976. (c) The heart and dart moth series —— including and ——— excluding
site 308.

12.3(b) are similar to estimates of relative national population sizes for these species from 1968 to 1984, also plotted on a logarithmic scale, but obtained from the same basic data by an entirely different method (Figure 5 and 22 in Woiwod and Dancey, 1987).

Series of index numbers, however computed, should always be plotted on a logarithmic scale. This has now been largely adopted; compare, for example, Pollard (1977) with Pollard (1982) or see Chapter 6 of this book. On the other hand, Ruger *et al.* (1986) still used linear plots for wildfowl index numbers.

A problem with the ratio method of calculating index numbers when the sites times years matrix is incomplete is that apparently temporary aberrations arising from the short-term loss or inclusion of prolific sites has a persistent effect on the series. This is very well illustrated by the heart and dart moth data, although it is an extreme example. Site number 308 contributed data only for the three years between 1976 and 1978; the number of heart and dart moths caught at that site in 1976 was exceptionally high compared with other sites, while in 1977 and 1978 it was well below the average. This results in the 1977/1976 ratio being less when this site is included than it would have been in its absence (0.82 against 1.35) and, accordingly, the 1977 index number is depressed. The important point is that the difference persists in the series from then onwards, even though site 308 makes no contribution to the data after 1978. The effect is shown in Figure 12.3(c). On a logarithmic scale, the inclusion of site 308 depresses the series by a constant amount from 1977 onwards. Thus index numbers can have their values affected by the temporary inclusion of a site, often largely a matter of chance, many years previously.

12.4 THE EFFECT ON INDEX NUMBERS OF TRANSFORMING RAW ABUNDANCE DATA

Site 308 was somewhat unusual in being exceptionally productive in 1976 (a very good year at most sites for the heart and dart moth), but also being significantly less productive in 1977 (which for most sites was an even better year). It has generally been found that variability in abundance, over time or space, is a function of a species' or a population's average abundance. The relationship, which has been found to hold for a wide range of species is known as Taylor's power law (Taylor 1961) and takes the form $s^2 = am^b$; s^2 and m are the variance and mean of the abundance data, and a and b are constants. Logarithmic transformation of the relationship yields $\log s^2 = \log a + b \log m$, i.e. the constants can be estimated by linear regression of $\log s^2$ against $\log m$. Figure 12.4(a) shows this for heart and dart moth abundances from 1965 to 1986, each point representing a year; all sites were used whether or not they contributed to paired-site data for index

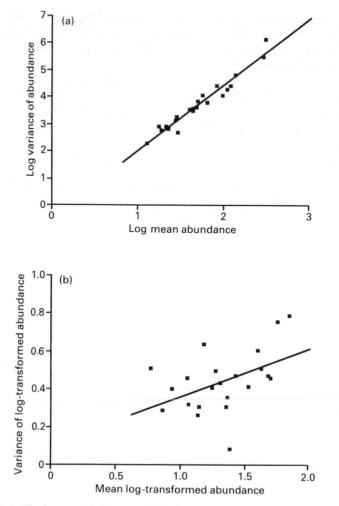

Figure 12.4 The heart and dart moth for the years 1965 to 1986. (a) Plot of log. variance against log. mean of untransformed site abundances. (b) Plot of variance against mean of logarithmically transformed site abundances.

numbers. The estimates are $\hat{a} = 0.37$ and $\hat{b} = 2.44$ ($t = 19.01$, d.f. $= 20$, $P < 0.001$); by comparison, Woiwod and Taylor (1984) quote $\hat{b} = 2.62$ for the years 1967 to 1982. The value of b can be used to indicate a suitable transformation for the data (Taylor 1961), e.g. square-root transformation when $b = 1$, logarithmic transformation when $b = 2$, or a negative fractional power when $b > 2$. The logarithmic transformation has been applied to raw site abundances, using $\log(x + 1)$ to deal with zero observations. The suitability of the transformation can be gauged by linear regression on untransformed axes of the variance and mean of the log. abundances, as in Figure 12.4(b). The variance is now only weakly dependent on the mean ($\hat{b} = 0.25$; $t = 2.37$, d.f. $= 20$, $P < 0.05$). This analysis has concerned the relationship over years of the mean and variance of abundance sampled across sites in each year. A corresponding analysis investigating the relationship over sites when abundance is sampled in different years for each site leads to similar conclusions, this time the variance becoming statistically independent of the mean after logarithmic transformation of the raw data.

It was suggested earlier, when dealing with a complete sites times years data matrix, that index numbers based on the sums of log.-transformed abundances held some disadvantages compared with those computed from geometric means, i.e. taking the antilog. of the arithmetic average of log.-transformed abundances. Figure 12.5(a) shows the series of index numbers for the heart and dart moth using paired-site sums of log.-transformed abundances. The series is much damped and only for two years does the index number fall below 100. The difference betwen two analyses, including or excluding site 308, is so small that it is not shown in Figure 12.5(a) as it would be hardly visible. The series using ratios of geometric means of paired sites is shown in Figure 12.5(b) and makes interesting comparison with Figure 12.3(c). The general shape is similar, except that index numbers are higher, particularly so in later years. The effect of excluding site 308 is detectable, but small. Note how the shape of the series in 1976 to 1978, with or without site 308, is similar to that in Figure 12.3(c) without site 308 – a significant improvement.

The question arises as to whether the heart and dart moth really has been relatively rare in 1979 to 1983 and in 1986, as suggested by Figure 12.3(c), or not, with the possible exception of 1981, as suggested by Figures 12.5(a) and 12.5(b). This is difficult to answer from the available published data. It is worth noting, however, that the paired-site ratio method of calculating index numbers does not use all the information available for a given year; only those sites which were also included in the previous year. For example, there are abundance data for 88 sites in 1980, whereas only 63 of these provided data in 1979 and are used, therefore, to calculate the 1980 index number. Hopefully, a year's index number should also reflect general abundance in this larger sample of sites. The mean abundances of *all* sites are illustrated in Figure 12.5(c), plotted on a logarithmic scale to aid comparison with Figures 12.3(c), 12.5(a) and 12.5(b). It is clear that, with the exception of 1981, mean abundances were not notably low in 1979 to 1986; Figure 12.5(c) agrees quite well with Figure 12.5(b) where index numbers were based on geometric means.

Some insight into the problem can be gained by plotting for each year index numbers (I) against mean abundances (m) for all sites that returned data. Ideally, the index number should be directly proportional to mean abundance, i.e. $I = am^b$ where a is a constant and $b = 1$. It follows that

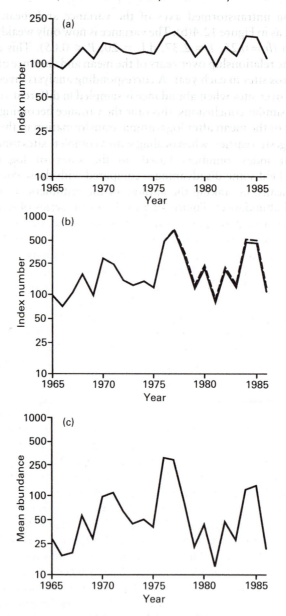

Figure 12.5 (a) and (b) Heart and dart moth index numbers (a) using $\log(x+1)$ transformed site abundances and (b) using geometric means of site abundances; —— including and − − − excluding site 308. (c) Mean abundance of the heart and dart moth for all sites irrespective of whether or not they were used in calculating index numbers.

$\log I = \log a + b \log m$, so that a plot of $\log I$ against $\log m$ should yield a straight line with slope $= 1$ and intercept $= \log a$. This plot is shown in Figure 12.6(a) for index numbers derived from untransformed data (i.e. those of Table 12.8 using 1965 as the reference base). The points clearly separate into three distinct lines. For 1965 to 1968 $\log I = 0.86 + 0.77 \log m$ ($t = 15.73$, d.f. $= 2$, $P < 0.01$), for 1969 to 1976 $\log I = 0.39 + 0.91 \log m$

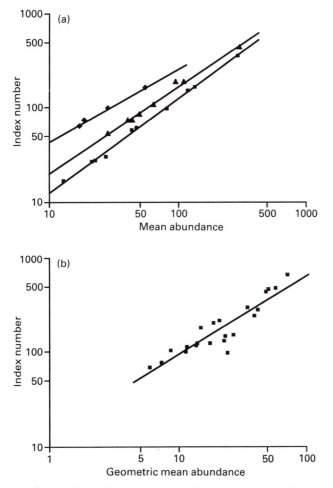

Figure 12.6 Index numbers for the heart and dart moth as a function of mean abundance for all sites whether or not they were used in calculating index numbers. (a) Index numbers obtained from untransformed paired-site abundances; ◆ 1965 to 1968, ▲ 1969 to 1976, ■ 1977 to 1986. (b) Index numbers obtained from geometric means against geometric mean abundances.

(t = 24.13,   d.f. = 6, $P < 0.001$),   and   finally   for   1977   to   1986 $\log I$ = 0.11 + 0.99 $\log m$ (t = 55.88, d.f. = 8, $P < 0.001$). These tests of significance are for the null hypothesis that the slopes are zero; the high values for t reflect the very close fits of the points to their respective lines. When the slopes are tested for agreement with the predicted value of one, that for 1965 to 1968 differs (t = 4.61, d.f. = 2, $P < 0.05$) whereas the other two do not (t = 2.30, d.f. = 6, $P > 0.05$ for 1969 to 1976 and t = 0.38, d.f. = 8, $P > 0.6$ for 1977 to 1986). It would appear that the relationship holds good for a number of years but then lurches onto a new alignment, this having happened twice. The severity of the lurch can be best judged by comparing the values of a, i.e. 7.2 (1965–8), 2.5 (1969–76) and 1.3 (1977–86). Note, however, that because the fitted lines converge slightly the effect is greatest on low index numbers. It implies that at the start of the series the index numbers were certainly much higher, relative to the mean abundance at all sites, than they were at the end, i.e. in the 1980s a falsely pessimistic impression is gained.

When index numbers calculated from geometric means of paired sites are plotted against geometric mean abundance of all sites for that year, as in Figure 12.6(b), a somewhat different picture emerges. Points no longer fall firmly onto clearly different lines; instead they are associated rather more loosely with a single line where $\log I$ = 1.14 + 0.84 $\log m$ (t = 9.82, d.f. = 20, $P < 0.001$ for slope = 0 and t = 1.89, d.f. = 20, $P > 0.05$ for slope = 1).

12.5 CONCLUSIONS

The construction of index numbers is undoubtedly a useful technique for summarising and presenting trends in wildlife monitoring data. Their behaviour is relatively easy to investigate when datasets are complete. Species' abundances differ between sites, as also does variation in abundance within a site from one time to another. Combination of data from different sites into truly representative index numbers may not be best achieved by simple summation across sites of the raw abundance data. The use of geometric means would seem to have some advantages; the more prolific sites, which are often the more variable, do not dominate index numbers. In particular, different series of index numbers can be combined into a composite series that is independent of the reference bases chosen for the constituent series.

The situation with incomplete datasets is less clear and large-scale monitoring schemes are likely to fall into this category. A year-to-year paired-site ratio method has been the favoured approach. It is clear, however, that a series of index numbers generated by this method can be affected on a long-term basis by the inclusion of a short run of atypical data, and that the series can occasionally adopt a new relationship relative to actual abundances, thus making comparisons inaccurate over longer periods of time. The example data have suggested, again, that the use of geometric means may be advantageous. It must be stressed that this need not be the case for all datasets, although analysis of the magpie moth data, not presented here, does show similar features. Further investigation is required. A way to proceed could be to take a large complete sites times years dataset and use a computer to generate repeatedly random 'holes' in the data matrix. The distributions of series calculated by different methods could then be assessed for their relative fits to the series based on the complete data.

There is undoubtedly a growing general familiarity with the concept of index numbers because of the increasing impact of those reflecting the nation's economic performance on day-to-day issues, for example wage-bargaining, pensions and the wider participation in stocks and shares. There is the danger, therefore, that an index number may be perceived as having an authority that is not justified by the quality of the original data on which it is based or by the method of its calculation. It is important to aim for the best methods. On the other hand, seemingly complex statistical manipulations and transformations may sow the seeds of mistrust in the minds of some.

Remember that even the RPI is the subject of on-going controversy, research and occasional refinement. There is an element of art, as well as science, in the construction of index numbers.

ACKNOWLEDGEMENT

I am grateful to Dr Michael Usher for his comments on this chapter and to the organisers and volunteers of the Rothamsted Insect Survey whose published data I have used.

REFERENCES

Department of Employment and Productivity (1967) *Method of Construction and Calculation of the Index of Retail Prices*, HMSO, London.

Fry, V. and Pashardes, P. (1986) *The Retail Prices Index and the Cost of Living*, The Institute for Fiscal Studies, London.

Mather, K. and Jinks, J.L. (1982) *Biometrical Genetics* (3rd edn), Chapman and Hall, London.

Pollard, E. (1977) A method for assessing changes in the abundance of butterflies, *Biological Conservation*, **12**, 115–34.

Pollard, E. (1982) Monitoring butterfly abundance in relation to the management of a nature reserve, *Biological Conservation*, **24**, 317–28.

Ruger, A., Prentice, C. and Owen, M. (1986) *Results of the IWRB International Waterfowl Census 1967–1983*, International Waterfowl Research Bureau, Slimbridge.

Taylor, L.R. (1961) Aggregation, variance and the mean, *Nature*, **189**, 732–5.

Taylor, L.R. (1986) Synoptic dynamics, migration and the Rothamsted Insect Survey, *Journal of Animal Ecology*, **55**, 1–38.

Taylor, L.R., French, R.A., Woiwod, I.P., Dupuch, M.J. and Nicklew, J. (1981) Synoptic monitoring for migrant insect pests in Great Britain and Western Europe. I. Establishing expected values for species content, population stability and phenology of aphids and moths, *Rothamsted Experimental Station. Annual Report for 1980*, **Part 2**, 41–104.

Williamson, M. (1972) *The Analysis of Biological Populations*, Edward Arnold, London.

Woiwod, I.P. and Dancy, K.J. (1987) Synoptic monitoring for migrant insect pests in Great Britain and Western Europe. VII Annual population fluctuations of macrolepidoptera over Great Britain for 17 years, *Rothamsted Experimental Station. Annual Report for 1986*, **Part 2**, 237–64.

Woiwod, I.P. and Taylor, L.R. (1984) Synoptic monitoring for migrant insect pests in Great Britain and Western Europe. V. Analytical tables for the spatial and temporal population parameters of aphids and moths, *Rothamsted Experimental Station. Annual Report for 1983*, **Part 2**, 261–93.

Biological monitoring and ecological prediction: from nature reserve management to national state of the environment indicators

PAUL A. KEDDY

13.1 INTRODUCTION

The techniques for biological monitoring are well-developed, as shown by examples in the preceding chapters. This chapter will explore two further issues: firstly, how do we decide what to measure, and secondly, what do we do with the data once it is collected? By examining these questions, we may make modifications to our biological monitoring programmes and growing computer databases. To restate, the issues which I propose to address are the following.

1. The selection of the state variables to be monitored. With more than a million species on the planet (Myers 1985; May 1989) we cannot possibly monitor each. Moreover, given predictions that a quarter may be extinct by the end of the next century we cannot possibly monitor even the threatened ones. How do we choose what to monitor? I will suggest a greater emphasis be placed upon macro scale state variables.
2. The interaction between monitoring and decision making. Monitoring is by its very nature *post hoc* – that is, it can only tell us what has already happened. But to wisely manage the biosphere, we need to predict future events. What is the relationship between biological monitoring, prediction, and decision making?

13.2 CHOOSING THE STATE VARIABLES TO MONITOR

13.2.1 Scale consideration

What do we monitor? Our first response is often to select a target species and begin to record its abundance through time. Target species can be selected for three reasons.

1. The species itself is of interest and we wish to measure our success in managing for it (e.g. Giant Pandas, Reid 1989).
2. The species is undesirable, and we wish to monitor our success in eradicating it (e.g. Purple Loosestrife in Canada, Thompson *et al.* 1987).
3. The species is an indicator of environmental conditions, and we are monitoring it because it is an indicator of the state of health of a particular habitat or environment (e.g. Meadow Beauty, Keddy *et al.* 1989).

Since the planet has millions of species (Myers 1985; May 1989), it seems that in most cases a species by species approach will leave the biologists outnumbered by species to be monitored. This suggests we need to place emphasis upon indicators of the state of health of entire ecosystems (Rapport *et al.* 1981; Ryder and Edwards 1985; Stokes and Piekarz 1987; Rapport 1990).

To give a concrete example, and one different from those discussed elsewhere, consider the problems in conserving the Atlantic coastal plain flora on the east coast of North America (Keddy 1985; Keddy and Wisheu 1989a; Wisheu and Keddy 1989a). What should be monitored to measure our success (or lack thereof) of conservation efforts in this region? A species by species approach would be time-consuming, and in the Canadian context, there are neither the trained staff, nor the financial resources to do this properly. However, a single species such as the showy pink-flowered Plymouth Gentian (*Sabatia kennedyana*) indicates species-rich coastal plain communities (Keddy 1985). Monitoring the abundance of this species may therefore measure the state of health of coastal plain communities in general.

But perhaps there is something even easier to measure than the abundance of Plymouth Gentians. We know that there are several environmental variables which can be used to predict the distribution of these species: naturally fluctuating water levels and sand shores (Keddy and Wisheu 1989a). Cottage development and all-terrain vehicles are key threats. Perhaps simply the number of lakes without dams and with less than ten cottages on them would be a simpler index still, and this sort of data might be collected by satellite or aerial photographs.

The choice of state variable will vary with the scale of the problem. At one end of the scale we may well be interested in monitoring the abundance of Plymouth Gentians in a single nature reserve. But at the other end of the scale, we may be interested in monitoring the state of health of Canadian wetlands as a whole, or the state of health of our entire natural environment, in which case no single state variable will suffice. In the case of wetlands, we would need to consider a hierarchy of scales (Table 13.1); each might require a different state variable to monitor its status.

As a guide to selecting appropriate state variables, Rapport (1990) suggests that in general we can recognise nine classes of information which provide symptoms of ecosystem distress: abnormalities, primary productivity,

Table 13.1 The choice of state variable for monitoring requires consideration of geographic and biological scales. In the case of wetlands, we might use a different state variable for each of the following scales of interest

---

Geographic scale
  global
  biogeographic region
  country
  state/county
  watershed
  individual nature reserve

Biological scale
  all wetlands
  specific wetland habitat
  functional group of organisms
  selected species

---

nutrients, species diversity, instability, disease prevalence, size spectra and contaminants. The challenge will be to provide simple means to measure these and to provide clear standards which must be achieved to maintain ecosystem health.

13.2.2 The Canadian situation

In Canada there is a large land base and small population size, exactly the opposite situation from the UK and Europe. Canadians therefore have had to think carefully about what should be monitored and at what scale (e.g. Stokes and Piekarz 1987; Bird and Rapport 1986). Environment Canada now has a State of Environment Reporting Branch to deal with biological monitoring at the national scale. Environmental monitoring is a category somewhat broader than biological monitoring, and includes three sets of environmental data (Stakeholder Group on Environmental Reporting 1987):

1. environmental assets (e.g. natural resource stocks);
2. agents of environmental change (e.g. resource harvest and depletion);
3. environmental quality (e.g. levels and trends of pollutants).

An alternative framework for the development of environment statistics has four categories (United Nations 1989):

1. Social and economic activities, natural events (e.g. use of natural resources, waste loadings).
2. Environmental impacts of activities/events (e.g. resource depletion, change in environmental quality).

3. Responses to environmental impacts (e.g. pollution control, resource management).
4. Stocks, inventories and background conditions (e.g. biological resources, energy stocks).

Table 13.2 Environmental databases available for State of Environment reporting in Canada (from Keddy and McRae 1989)

Category	No. of databases
I. Environmental assets	
Water resources	3
Land resources	5
Wildlife	7
General	7
II. Agents of environmental change	
Waste generation and disposal	
Mining and smelting	3
Manufacturing	4
Municipal	2
Transportation	1
PCBs	3
Stored waste	1
Ocean dumping	2
Emissions – air	2
Environmental restructuring	
Urbanisation	1
Harvesting	
Agriculture	1
Hunting and trapping	1
Pesticide applications	1
General	3
III. Environmental quality	
Air	1
Water	3
Wildlife populations	
(plants and animals)	3
Landscape	1
Perceptions	1
General	4
IV. Other (indirectly related to environmental evaluation)	8

The document cited above lists a comprehensive series of indicators for each of these four categories.

There are growing numbers of databases which include the data for environmental monitoring. Friend (1988) listed 32 database categories suitable for this task. They include biological information on variables such as fish harvests, population sizes of migratory birds, and pesticide residues in organisms. A more recent inventory for the State of Environment Reporting Branch (Keddy and McRae 1989) identified 55 databases within Environment Canada alone. These range (Table 13.2) from avian census plots (non-game bird counts on permanent plots in various habitat types across Canada) to the national air pollution surveillance network (continuous monitoring of common air contaminants).

The following examples illustrate some of the biological variables being monitored in Canada, with emphasis upon larger scale biological variables.

13.2.3 Ducks and wetlands

Environment Canada has a long history of interest in waterfowl, in part because of the hunting lobby, and in part because of Canada's obligations

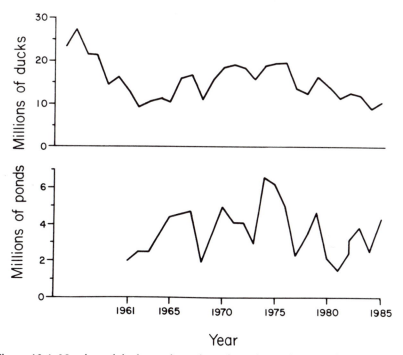

Figure 13.1 Number of ducks, and number of ponds, on the Canadian prairies for approximately the last 25 years (Bird and Rapport 1986).

under the Migratory Birds Convention Act. Figure 13.1 shows the number of ducks in the Canadian prairies as a function of time. There is clearly a problem. These data are particularly interesting, however, because we can consider ducks as bioindicators of wetlands. These data therefore may tell us of the declining state of health of prairie wetlands (Figure 13.1, bottom). We may be far more interested in the state of health of wetlands as a whole than any single group of species in them.

One could argue that the number of ducks is a poor choice to measure health of prairie wetlands – it may be much simpler to measure acreage of wetlands directly from satellite photographs. But an advantage to monitoring ducks is that they may be integrating both the quantity and the quality of wetlands. Increasing pesticide use may be reducing wetland quality (Sheehan *et al.* 1987). On the other hand, the decline in duck populations might also be caused by overhunting in the US – in which case the duck population sizes are telling us little about the state of health of prairie wetlands. In fact, the data probably integrate the effects of declining area of wetland, declining quality of wetland, and increased hunting pressure as well as many other unknown factors.

13.2.4 Great Lakes fisheries

Changes in the Canadian Great Lakes illustrate the scale and rate at which freshwater ecosystems are being altered by human activities. Figure 13.2

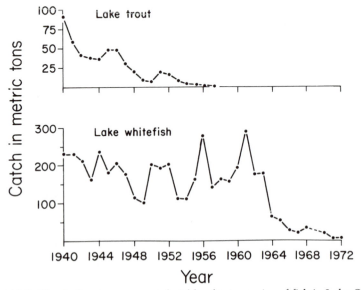

Figure 13.2 The decline in commercial yields of two species of fish in Lake Ontario since 1940 (after Christie 1974).

shows declines of two important commercial species over time. Christie (1974) has discussed such changes, and has attributed them to human impacts including overfishing, eutrophication, dams and the introduction of exotic species. We might ask, however, whether there are indicators of the state of health of Great Lakes fisheries beyond species by species enumerations. Figure 13.3 shows two indicators which illustrate overall changes in fish community structure over the last century. The top panel shows that large benthic species such as lake trout and sturgeon

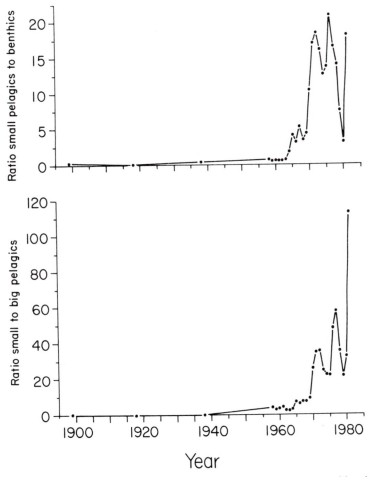

Figure 13.3 Changes in the state of Great Lakes ecosystems as measured by changes in composition of Lake Ontario fish communities. Small pelagics include smelts and alewife; benthics include lake whitefish and sturgeon; big pelagics include blue pike and lake herring (after Rapport 1983, Table 6).

predominated early in this century, but they have now been replaced by small pelagics such as smelts and alewife. The bottom panel shows that pelagic species have been shifting in composition; early in the century, big pelagic species such as blue pike and lake herring were most common; now they have been replaced by small pelagics such as smelt and alewife. While such indicators do not provide information on single species, they may be far better for monitoring the overall state of health of freshwater ecosystems. Regier *et al.* (1988) and Rapport (1989) have discussed the extent to which similar patterns are found in the Baltic Sea. More generally, the tendency for stress to eliminate larger species seems to be a common feature of ecosystem distress (Rapport *et al.* 1985; Rapport 1989, 1990).

13.2.5 Endangered species

There are approximately 5092 species of plants and vertebrate animals known from Canada (Pollard and McKechnie 1986). Of these, 95 species or subspecies are considered threatened, endangered, extirpated or extinct by the Committee on the Status of Endangered Wildlife in Canada (Table 13.3) – approximately 2% of our biota. A simple biological measure of environmental health would be the number of species falling into these categories (Figure 13.4). At present, a plot of number of species in this category against time would not be a reliable way of monitoring biological health. First, invertebrates are conspicuously absent. Second, many species are not listed by COSEWIC simply because there is little money to support the work needed to prepare the reports to have species officially listed! That is, the number of such species is currently underestimated for financial reasons,

Table 13.3 Monitoring threatened wildlife in Canada. Status designations by the Committee on the Status of Endangered Wildlife in Canada as of April 1989

Category (status)	Birds	Mammals (terrestrial)	Mammals (marine)	Fish	Plants	Amphibians and reptiles	TOTAL
Rare	16	13	3	22	18	1	73
Vulnerable	0	1	1	8	1	2	13
Threatened	7	5	2	10	18	–	42
Endangered	8	3	5	3	17	1	37
Extirpated	–	2	2	2	1	–	7
Extinct	3	1	1	4	–	–	9
TOTAL	34	25	14	49	55	4	181

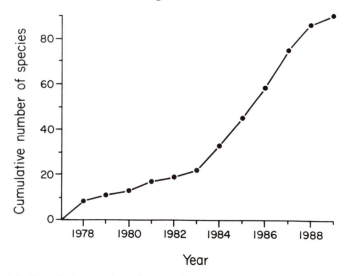

Figure 13.4 Cumulative number of species designated by COSEWIC (Committee on the Status of Endangered Wildlife in Canada) as threatened, endangered, extirpated or extinct in Canada. 'Species' can mean subspecies or geographically restricted population in COSEWIC data, but where a species was designated more than once in such categories, it was counted only once for this figure. Only plants and vertebrate animals are currently considered by COSEWIC.

and changes with time at present reflect the number of reports completed per year rather than the number of species being newly-threatened. Once this lag phase is finished, however, this would be a useful indicator. An allocation of a few hundred thousand dollars would probably be sufficient to finish candidate species (excluding invertebrates). Two species, the White Pelican and Wood Bison were recently downlisted from the endangered category, showing that percent of fauna listed by COSEWIC could also measure the success of recovery programs.

13.2.6 Protected land base

An ultimate cause of endangerment is generally habitat destruction (Ehrlich and Ehrlich 1981). A simple measure of our success in protecting habitat would have two components: the total area of land protected, and the degree to which this land base represents the natural diversity of Canada. In spite of our small population size and large area, Canada lags behind many countries in terms of total protected area in our parks system (Task Force on Park Establishment 1987). Parks Canada recognises 39 terrestrial and 29 marine

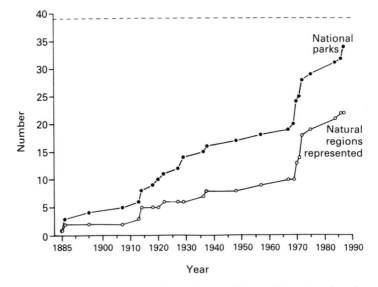

Figure 13.5 Monitoring the state of completion of Canada's national park system. The hatched line at the top represents the goal – complete representation of national park natural regions in Canada. While the number of national parks has slowly grown to 34 (solid dots), only 22 natural regions occur in a park (open dots). Moreover, some of these 22 natural regions are only partially represented, so that the parks system currently is less than half complete.

regions in the country, and Figure 13.5 shows that many of these important natural regions are not yet protected by National Parks (Taschereau 1985; Task Force on Park Establishment 1987). The World Wide Fund for Nature has just launched its national Endangered Spaces programme (Hummel 1989) which provides specific targets for each province to meet in order to ensure a national system of representative protected areas.

13.3 BIOLOGICAL MONITORING AT THE NATIONAL SCALE: TOWARDS STATE OF ENVIRONMENT INDICATORS FROM THE WORLD CONSERVATION STRATEGY

We need indicators of the biological health of a country in much the same way that the GNP of the country provides one measure of its economic health. No one indicator is sufficient, of course; even the GNP is supplemented by measures such as foreign exchange rates and un-employment rates. In the absence of measurable criteria for environmental improvement, we will have little way to judge whether our increasing expenditures on environment are really paying off. That is, are they really leading to measurable improvement in the state of environment? And at a time when political parties are rushing to paint themselves green, indicators such as this would be a simple way for the public to tell whether a party's policies once elected matched its rhetoric. There is general agreement about neither the components of such measures nor their relative weightings (e.g. Inhaber 1974; Stokes and Piekarz 1987; Karr 1987; Liverman *et al.* 1988). One approach which might be considered is indicators based upon the World Conservation Strategy (International Union for Conservation of Nature and Natural Resources 1980). The World Conservation Strategy gives us three components of biological health of the planet: genetic diversity, sustainable utilisation and the maintenance of essential life support systems (Figure 13.6). Using indicators ('state variables') for these three components would provide relatively simple measures of environmental health that could be monitored annually and reported to a country or province without it. It might also be possible to combine these indicators into more general indices summarising one aspect of the state of environmental health, such as bio-diversity (Figure 13.6). Other possible indicators are presented in a recent review prepared for the United Nations (1989).

The state variables in Figure 13.6 were chosen because of their obvious value as indicators, and because the data are already available (Friend 1988; Keddy and McRae 1989), or could be collected relatively cheaply. We could therefore, if we wished, have biological monitoring of each nation, and each political unit within a nation, in relatively short order. These indicators and/ or indices could be compiled by a non-profit organisation (such as the World Wide Fund for Nature). It might be best for government to delegate such responsibilities to avoid any suggestion of political interference in the annual

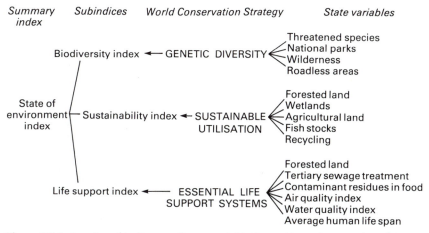

Figure 13.6 A series of indicators ('state variables') can be used to monitor the state of a nation's health according to the World Conservation Strategy. These indicators would provide quantitative measures of the success of environmental programs. It might also be possible to combine indicators into summary indices.

environmental audit, or it might be set up with the same independence as the Canadian Auditor General who evaluates federal expenditures for efficiency.

13.4 PREDICTION, MONITORING AND DECISION MAKING

13.4.1 Prediction

What is the relationship between monitoring and prediction and why does it matter? There are two reasons why prediction is important. First, it is an essential component in the maturation of ecology as a science (Rigler 1982; Peters 1980a,b). Second, and of greater importance for conservation, decisions we make today are guided by our expectations, or predictions, about their consequences in the future. To return to the economic analogy, accurate predictions are necessary to guide political decision making. If we increase personal income taxes on low income tax payers, what will be the change in the national seasonally-adjusted unemployment rate? As ecologists, we need to be able to make similar predictions. If the James Bay hydroelectric project (Phase 2) goes ahead, what will be the impact on coastal ecosystems? If Montreal spends money to install tertiary sewage treatment, how will aquatic communities downstream respond? If the climate warms by 1°C, how much marginal agricultural land in the prairies should be converted to pasture or abandoned? If Canada's remaining coastal forests are clear cut, what will happen to oceanic fisheries?

At present we are unable to make such predictions. This may have less to do with inherent ecosystem characteristics than with human behaviour: the emphasis in much of ecology is still upon description rather than prediction, perhaps because the foundations of ecology are still closely tied with natural history. It is beyond the scope of this chapter to explore in detail the different avenues which have been proposed for prediction, but the issue is explored further in Holling (1978), Peters (1980a,b), Rigler (1982), Starfield and Bleloch (1986) and Keddy (1989). A few generalisations are possible, however.

In the simplest case, making predictions requires that we decide what we want to predict (the dependent variable), and then determine what independent variable will be the best predictor of the dependent variable. This approach raises important questions. First, what are the key state variables that we need to predict? One of our scientific problems is that we do not yet have a body of ecological models to tell us which variables are the key state variables for maintaining the state of health of our planet or measuring ecosystem 'integrity'. To use a medical analogy (Rapport *et al.* 1979; see also Rapport *et al.* 1981; Rapport 1990), we do not yet know whether to measure blood pressure, hair colour, weight, or smoking habits to predict life span. Lewontin (1974) has referred to this as 'the agony of community ecology'.

Similarly, we have few guides as to what state variables will be the best predictors of our independent variables. The best predictor of future endangered species lists may be percent of ecoregions represented in national parks rather than simply percent of Canada's area devoted to parks. Working out the empirical relationships among such state variables would allow us to make such predictions (e.g. Rigler 1982).

We may find that in general, higher order (or macro level) variables will be easier to predict. To borrow an example from physics, the behaviour of individual molecules is very difficult to predict, but the ideal gas laws nonetheless provide clear empirical relationships among pressure, volume, and temperature. Prigogine and Stengers (1984) provide many examples of such relationships in emergent properties, as the title of their book, *Order out of Chaos*, suggests. I have argued elsewhere that this applies directly to community ecology (Keddy 1987).

Emphasising higher order state variables has another advantage: it tends to highlight general properties rather than site specific ones. This is necessary because there are insufficient resources for dealing with conservation issues on a case by case basis. Leary (1985) has discussed the importance of this transition, and presented a general model as a guide in the evaluation of government research (Figure 13.7).

A good example of this approach to prediction is the relationship between species richness (the dependent variable) and standing crop (the independent variable) in vegetation. Early work (Al-Mufti *et al.* 1977; Grime 1979) demonstrated an empirical relationship between these variables, with species richness reaching a maximum at intermediate levels of biomass. Since then similar relationships have been found in vegetation types including fens (Wheeler and Giller 1982), Pine savannas (Walker and Peet 1983) and

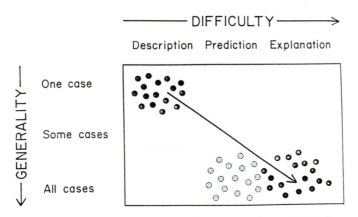

Figure 13.7 Two criteria for judging scientific progress (Leary 1985). Leary argues that we need to move from the upper left to the bottom right. For most conservation management, prediction (grey dots) would be the first priority (from Keddy and Wisheu 1989b).

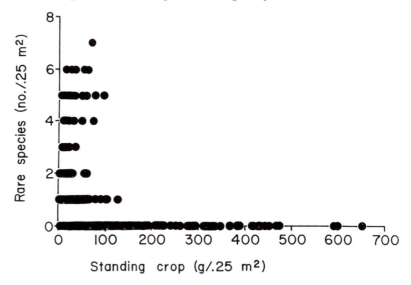

Figure 13.8 Vegetation biomass can be used to predict which wetland vegetation types do not support nationally rare wetland plant species. The number of nationally rare plant species in each of 401 quadrats from eastern Canada is plotted against quadrat biomass (dry weight). (Reproduced from Moore *et al.* (1989) with permission from Elsevier Publishers Ltd.)

freshwater shorelines (Wisheu and Keddy 1989b). From the point of view of conservation, this simple relationship has been valuable. Grime (1973) has discussed managing species-rich vegetation types in this context. Moore *et al.* (1989) show that biomass can be used to predict which wetlands will have nationally rare wetland plant species (Figure 13.8).

13.4.2 Monitoring and prediction

Monitoring is allied to prediction in two essential ways. First, in the model-building stage, it is only by making wrong predictions that we can build better models and monitoring is essential to determine whether our predictions were right or wrong. As in multiple regression techniques (e.g. Kleinbaum and Kupper 1978; Sokal and Rohlf 1981), adding in new independent variables will gradually increase the accuracy of predictions. Of course, no ecological model can be expected to produce prediction, so monitoring plays the more important role of providing an early warning system to advise us that the real world is beginning to deviate from our expectations. Consider species richness again. We may well know that to achieve maximum richness of chalk grassland, we need vegetation with biomass levels under $500\,\mathrm{g\,m}^{-2}$.

We therefore plan grazing regimes accordingly. But if air pollution increases soil fertility, as is happening over much of Europe (Ellenberg 1988), higher than predicted levels of grazing may be needed to maintain the desired vegetation biomass. Furthermore, even if we maintained vegetation biomass constant, there might still be a shift towards species more tolerant of eutrophication (Ellenberg 1988), requiring us to devise yet another management strategy for our grassland. Monitoring therefore provides constant feedback to managers, and constant feedback for refining and reconstructing predictive models. Any political decision with an environmental component has an element of prediction, even if the simplest (and commonest) 'prediction' in the past has been that nothing of any consequence will be changed by the decision. With a growing environmental awareness by politicians and industrialists, there will be a growing interest in accurate predictions, or at least possible scenarios, to guide decision making.

13.4.3 Monitoring and decision making

Monitoring is the essential feedback link between humans and their environment. Once a decision is made, monitoring provides feedback about the consequences of that decision. It is therefore essential that there be well-established mechanisms for ensuring that the information gained from monitoring leads to modifications or even reversals of earlier decisions. If the right-hand arrow in Figure 13.9 is missing, we are losing the most important part of monitoring in both the scientific and political arenas.

The importance of feedback from monitoring applies at the complete range of scales as well. At the most local scale, the decision of a nature reserve manager may be to fluctuate water levels over a ten-year cycle to produce rich coastal plain vegetation in a lake (left arrow) (e.g. Keddy *et al.* 1989). The nature reserve management plan should then have a requirement that

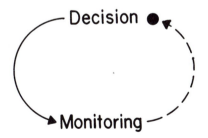

Figure 13.9 Monitoring records the consequences of environmental decisions. It is therefore important to have a feedback loop (hatched arrow) so that observations from monitoring can produce changes in social decisions as early as possible.

monitoring be carried out so that the merits of that decision can be evaluated, and the water level regime modified at a later date. At the large scale, the decision by the Quebec government to flood large areas of northern Quebec to generate hydroelectric power for the United States (the James Bay Project Phase 2) has the left-hand arrow, but the right-hand one is absent. There is no mechanism for removing the dams and restoring the environment if environmental quality deteriorates. Adaptive impact assessment (Holling 1978) and sustainable development will require such a feedback loop.

We can expect that in the future all environmental decisions, be they the management of a nature reserve or a national economy, will require environmental auditing both to guide initial decisions and to monitor projects as they progress so that modification is possible. Choosing the biological variables to monitor, and building the linkages and feedback loops to ensure that they influence societal behaviour, are two challenges which face us. As we address these two issues, biological monitoring will become an increasingly powerful tool for decision making at local, regional and national scales.

ACKNOWLEDGEMENTS

Many thanks to Brent Coates, Tony Friend, Connie Gaudet, Cathy Keddy, and Dave Rapport for providing information to help complete this chapter. They bear no responsibility for my interpretation of the information they provided. I also thank Anita Payne and Irene Wisheu for their help in preparing the manuscript, and Barrie Goldsmith for his helpful editorial comments.

REFERENCES

Al-Mufti, M.M., Sydes, C.L., Furness, S.B., Grime, J.P., and Band, S.R. (1977) A quantitative analysis of shoot phenology and dominance in herbaceous vegetation, *Journal of Ecology*, **65**, 759–91.

Bird, P.M. and Rapport, D.J. (1986) *State of the Environment Report for Canada*, Environment Canada, Ottawa.

Christie, W.J. (1974) Changes in the fish species composition of the Great Lakes, *Journal of the Fisheries Research Board of Canada*, **31**, 827–54.

Ehrlich, P. and Ehrlich, A. (1981) *Extinction*, Random House, New York.

Ellenberg, H. (1988) Floristic changes due to nitrogen deposition in Central Europe, in *Critical Loads for Sulphur and Nitrogen* (eds J. Nilsson and P. Grennfelt), Report from a workshop held at Skokloster, Sweden, 19–24 March 1988. Nordic Council of Ministers.

Friend, A.M. (1988) Federal government databases relevant for environmental risk management, in *Information Needs for Risk Management* (eds C.D. Fowle, A.P. Grima and R.E. Munn), Institute for Environmental Studies, University of Toronto, Toronto, pp. 63–89.

Grime, J.P. (1973) Control of species density in herbaceous vegetation. *Journal of Environmental Management*, 1, 151–67.

Grime, J.P. (1979) *Plant Strategies and Vegetation Processes*, Wiley, Chichester.

Holling, C.S. (ed.) (1978) *Adaptive Environmental Assessment and Management.* Wiley, Chichester.

Hummel, M. (1989) *Endangered Spaces: The Future for Canada's Wilderness*, Key Porter Books, Toronto.

Inhaber, H. (1974) Environmental quality: outline for a national index for Canada, *Science*, 186, 798–805.

International Union for Conservation of Nature and Natural Resources (1980) *World Conservation Strategy*, IUCN, UNEP and WWF.

Karr, J.R. (1987) Biological monitoring and environmental assessment: a conceptual framework, *Environmental Management*, 11, 249–56.

Keddy, C.J. and McRae, T. (1989) *Environmental Databases for State of the Environment Reporting: Conservation and Protection Headquarters*, Report No. 9, Strategies and Scientific Methods, SOE Reporting Branch, Canadian Wildlife Service, Environment Canada.

Keddy, P.A. (1985) Lakeshores in the Tusket River Valley, Nova Scotia: distribution and status of some rare species, including *Coreopsis rosea* Nutt. and *Sabatia kennedyana* Fern. *Rhodora*, 851, 309–20.

Keddy, P.A. (1987) Beyond reductionism and scholasticism in plant community ecology, *Vegetatio*, 69, 209–11.

Keddy, P.A. (1989) *Competition*, Chapman and Hall, London.

Keddy, P.A. and Wisheu, I.C. (1989a) Ecology, biogeography, and conservation of coastal plain plants: some general principles from the study of Nova Scotian wetlands, *Rhodora*, 91, 72–94.

Keddy, P.A. and Wisheu, I.C. (1989b) Why ignorance isn't bliss: biological considerations in wetlands management, in *Wetlands: Inertia or Momentum* (eds M.J. Bardecki and N. Patterson), Proceedings of a conference held in Toronto, Ontario, 21–22 October 1988. Federation of Ontario Naturalists.

Keddy, P.A., Wisheu, I.C., Shipley, B. and Gaudet, C. (1989) Seed banks and vegetation management for conservation: towards predictive community ecology, in *Ecology of Soil Seed Banks* (eds M.A. Leck, V.T. Parker and R.L. Simpson), Academic Press, San Diego.

Kleinbaum, D.G. and Kupper, L.L. (1978) *Applied Regression Analysis and Other Multivariate Methods*, Duxbury Press, Massachusetts.

Leary, R.A. (1985) A framework for assessing and rewarding a scientist's research productivity, *Scientometrics*, 7, 29–38.

Lewontin, R.C. (1974) *The Genetic Basis of Evolutionary Change*, Columbia University, New York.

Liverman, D.M., Hanson, M.E., Brown, B.J. and Merideth, R.W. (1988) Global sustainability: toward measurement, *Environmental Management*, 12, 133–43.

May, R.M. (1989) How many species are there on Earth? *Science*, 241, 1441–9.

Moore, D.R.J., Keddy, P.A., Gaudet, C.L. and Wisheu, I.C. (1989) Conservation of wetlands: do infertile wetlands deserve a higher priority? *Biological Conservation*, 47, 203–17.

Myers, N. (ed.) (1985) *The Gaia Atlas of Planet Management*, Pan Books, London.

Peters, R.H. (1980a) Useful concepts for predictive ecology. *Synthese*, 43, 257–69.

Peters, R.H. (1980b) From natural history to ecology. *Perspectives in Biology and Medicine*, **23**, 191–203.

Pollard, D.F.W. and McKechnie, M.R. (1986) *World Conservation Strategy – Canada: A Report on Achievements in Conservation*, Conservation and Protection, Environment Canada, Ottawa.

Prigogine, I. and Stengers, I. (1984) *Order Out of Chaos*, New Science Library, Shambala, Boulder, Colorado.

Rapport, D.J. (1983) The stress-response environmental statistical system and its applicability to the Laurentian Lower Great Lakes, *Statistical Journal of the United Nations ECE*, **1**, 377–405.

Rapport, D.J. (1989) Symptoms of pathology in the Gulf of Bothnia (Baltic Sea): ecosystem response to stress from human activity, *Biological Journal of the Linnean Society*, **37**, 33–49.

Rapport, D.J. (1990) What constitutes ecosystem health?, in *Perspectives in Biology and Medicine* (in press).

Rapport, D.J., Regier, J.A. and Hutchinson, T.C. (1985) Ecosystem behaviour under stress, *The American Naturalist*, **125**, 617–40.

Rapport, D.J., Regier, H.A. and Thorpe, C. (1981) Diagnosis, prognosis, and treatment of ecosystems under stress, in *Stress Effects on Natural Ecosystems* (eds G.W. Barrett and R. Rosenberg). Wiley, Chichester.

Rapport, D.J., Thorpe, C. and Regier, H.A. (1979) Ecosystem medicine. *Bull. Ecol. Soc. Amer.*, **60**, 180–2.

Reid, D. (1989) Panda Conservation Update, *Working for Wildlife*, Spring/Summer 1989, World Wide Fund for Nature.

Regier, H.A., Tuunainen, P., Russek, Z. and Persson, L. (1988) Rehabilitative redevelopment of the Fish and Fisheries of the Baltic Sea and the Great Lakes, *Ambio*, **17**, 121–30.

Rigler, F.H. (1982) Recognition of the possible: an advantage of empiricism in ecology, *Canadian Journal of Fisheries and Aquatic Sciences*, **39**, 1323–31.

Ryder, R.A. and Edwards, C.J. (eds) (1985) *A conceptual approach for the application of biological indicators of ecosystem quality in the Great Lakes basin*, International Joint Commission, Great Lakes Regional Office, Windsor, Ontario.

Sheehan, P.J. et al. (1987) *The Impact of Pesticides on the Ecology of Prairie Nesting Ducks*, Technical Report Series No. 19, Canadian Wildlife Service, Enviroment Canada.

Sokal, R.R. and Rohlf, F.J. (1981) *Biometry*, (2nd ed.), W.H. Freeman and Co., New York.

Stakeholder Group on Environmental Reporting (1987) *A Study of Environmental Reporting in Canada*, Environment Canada, Ottawa.

Starfield, A.M. and Bleloch, A.L. (1986) *Building Models for Conservation and Wildlife Management*, Macmillan, New York.

Stokes, P. and Piekarz, D. (eds) (1987) *Ecological Indicators of the State of the Environment*, Environmental Interpretation Division, Environment Canada, Ottawa.

Taschereau, P.M. (1985) *The Status of Ecological Reserves in Canada*, The Canadian Council on Ecological Areas and The Institute for Resource and Environmental Studies, Dalhousie University, Halifax.

Task Force on Park Establishment (Canada) (1987) *Our Parks – Vision for the 21st Century: Report to the Minister of the Environment*, Environment Canada, Parks and Heritage Resources Centre, University of Waterloo.

Thompson, D.Q., Stuckey, R.L. and Thompson, E.B. (1987) *Spread, Impact, and Control of Purple Loosestrife (*Lythrum salicaria*) in North American Wetlands*, United States Department of the Interior, Fish and Wildlife Service, Fish and Wildlife Research 2.

United Nations (1989) *Concepts and Methods of Environmental Statistics: Statistics of the natural environment – a technical report*, Studies in Methods, Statistical Office, Department of International Economic and Social Affairs, United Nations, New York.

Walker, J. and Peet, R.K. (1983) Composition and species diversity of pine-wiregrass savannas of the Green Swamp, North Carolina. *Vegetatio*, 55, 163–79.

Wheeler, B.D. and Giller, D.E. (1982) Species richness of herbaceous fen vegetation in Broadland, Norfolk in relation to the quantity of above-ground plant material, *Journal of Ecology*, 70, 179–200.

Wisheu, I.C. and Keddy, P.A. (1989a) The conservation and management of a threatened coastal plain plant community in Eastern North America (Nova Scotia, Canada), *Biological Conservation*, 48, 229–38.

Wisheu, I.C. and Keddy, P.A. (1989b) Species richness – standing crop relationships along four lakeshore gradients: constraints on the general model, *Canadian Journal of Botany*, 67, 1609–17.

Synthesis

BARRIE GOLDSMITH

Now that you have reached the end of this book I hope that you will have been sufficiently stimulated to want to embark upon a monitoring project but I hope that you will first pause and consider this checklist before starting.

1. Is there really a good reason for doing this? Are your objectives clear? Are you aiming to record global/local change (or both) or the effects of management or what?
2. Have your objectives been put in order/ranked?
3. Which physical, chemical and biological features are to be recorded?
4. How are you going to cope with taxonomic problems?
5. Would the use of aerial photography or remote sensing save time and money?
6. Are sampling problems complicated by time of year (season), stage of succession, or a cyclical event (such as burning)?
7. Have the sampling points been located?
8. Are you sure that they can be relocated by yourself/other people? Have they been marked in at least two ways?
9. What size sampling unit is to be used?
10. How many samples are needed (how much replication)? Has a statistician been consulted? Have you read Chapter 2?
11. If you are calculating index numbers have you read Chapter 12?
12. At what time interval will samples be taken (frequency), e.g. weekly, annually, every five years? Decide on the basis of periodicity/cycle of the event, expected rate of change, and the finance/time available (see Chapter 2).
13. At what time of the year will the monitoring take place? It may vary for different parameters/taxa/habitats.
14. Will monitoring be combined with experimentation, e.g. comparable recording in a grazing enclosure, manipulation of water-levels?
15. How will the data be analysed/presented? What statistical tests are appropriate? Is some kind of sensitivity analysis available (see Chapter 2)?

16. When will the monitoring stop? What criteria are there for reaching this stage? What happens then? See Chapter 2.
17. Who will administer/co-ordinate/file the results of the monitoring programme? A related question is who will decide on changes that might be needed?
18. Who will pay for it? Is the budget secure? Is the staffing organised? Might it be cheaper to buy a single piece of automatic equipment (e.g. an automatic gauging station) rather than having to pay a part-timer for several years (to read a stage-board)?
19. Have other scientists/local people seen your plans and been invited to comment on them/modify them?

Several themes appear throughout the preceding chapters. Perhaps the strongest is the plea that monitoring should only be carried out in relation to clearly stated objectives. John Hellawell emphasies that in the absence of such objectives a series of surveys becomes surveillance and not monitoring. Having identified one's objectives it is then necessary to sort them into a ranking of priorities. The criteria for doing this will vary from project to project but it will usually be necessary to separate top priority recording from intermediate and thirdly an optional but desirable category. In my Chapter 11 on monitoring overseas in an unfamiliar and relatively poorly documented location, the criteria used to select these categories were based on whether the biotopes and species were of international, national or regional importance. Biotopes and species of international importance had to be monitored as the top priority.

Having established our priorities the next step is to decide on the appropriate taxa for recording. They are usually those easiest to identify, i.e. birds, flowering plants or butterflies. However other taxa can be very sensitive to man's activities, for example, lichens are sensitive to pollution and aquatic invertebrates are used as indicators of water quality.

Techniques of data processing and statistical analysis need to be considered at the outset. Mike Usher spells this out in Chapter 2 but ecologists and conservationists so often overlook this subject in their headstrong rush to get out in the field and get on with it! Terry Crawford in Chapter 12 has considered the problems involved in formulating index numbers.

Monitoring data are of little value if they are not associated with information about concurrent management and environmental data. Peaks and troughs in fish data may be explained away as being due to fishing intensity which is usually expressed as over-fishing. However in lakes periods of flooding, pulses of pollution and agricultural or forestry activities in the catchment can all be important.

Monitoring, usually being a long-term activity, needs considerable organisation. It is not often the subject of funding by research council grants and is not ideal for PhD studies but it needs long-term commitment from the managing agency. Usually the costs of conducting good monitoring are considerable but some respected monitoring procedures are conducted on small budgets. Pollard's butterfly monitoring (Chapter 6) is a classic example which needs little more than the dedication of a weekly lunch hour, some taxonomic ability and the establishment of a co-ordinating agency to process the data.

When the data has been collected and the results processed it is then necessary to communicate them to those people who are in a position to effect the necessary response. This is easier said than done. To take an extreme example we now have data about the rate of burning of Amazonian forest but are failing to communicate that effectively to the Brazilian people, landowners or their regional or national governments.

Finally we have seen that all monitoring needs a cut-off or at least review built into it. Again the criteria for determining this need to be thought out very carefully. If this appears difficult it is likely to be because the objectives and priorities were not clearly thought out at the start of the exercise.

Index